Green Energy and Technology

More information about this series at http://www.springer.com/series/8059

Michael Gbolagade Oladokun
Clinton Ohis Aigbavboa

Simulation-Based Analysis of Energy and Carbon Emissions in the Housing Sector

A System Dynamics Approach

Springer

Michael Gbolagade Oladokun
Faculty of Engineering and the Built Environment
University of Johannesburg
Johannesburg
South Africa

Clinton Ohis Aigbavboa
Faculty of Engineering and the Built Environment
University of Johannesburg
Johannesburg
South Africa

ISSN 1865-3529 ISSN 1865-3537 (electronic)
Green Energy and Technology
ISBN 978-3-030-09207-8 ISBN 978-3-319-75346-1 (eBook)
https://doi.org/10.1007/978-3-319-75346-1

Printed on acid-free paper

This Springer imprint is published by the registered company Springer International Publishing AG part of Springer Nature
The registered company address is: Gewerbestrasse 11, 6330 Cham, Switzerland

Preface

The effects of climate change as a result of the rising profile of carbon emissions globally have forced the governments of many nations to collaborate their efforts in order to counteract the damage caused by greenhouse gas emissions. This has resulted in agreements by participating governments to reduce the effects of climate change, the latest of which was the declaration of the 2015 United Nations Climate Change Conference in Paris. As such, governments of nations who are signatories to the accord have initiated schemes targeting the reduction of carbon emissions in all spheres of their economies, including the housing sector. Hence, this book describes the development of a system dynamics-based model that can capture the future trajectories of housing energy and carbon emissions.

This research book views energy and carbon emissions in the housing sector as a complex sociotechnical problem involving the analysis of intrinsic interrelationships among the dwellings, occupants, and environment. Using the UK housing sector as illustration, the book demonstrates how the system dynamics simulation can be used as a learning laboratory regarding the future trends in housing energy and carbon emissions over time, especially until 2050. One of the unique features of the approach used in the book is that it adopts the pragmatist research strategy. This strategy involves the collection of both qualitative and quantitative data to develop the model, which is not well pronounced in traditional building physics and regression-based forecasting. Consequently, the book will improve construction researchers' understanding regarding the complexity involved in housing energy and carbon emissions from the systems thinking perspective, thereby extending the frontier of the knowledge base in system dynamics to energy issues in housing. Additionally, using a nondeterministic systems approach, the book generates further insights for the readers on future profiles of housing energy and carbon emissions.

The book will be of interest to researchers in the fields of architectural engineering, housing studies, and climate change, among others. This research book will also be of interest to industry practitioners and policy makers in housing energy. The authors affirm that the texts utilised in this book reflect original ideas and, where necessary, sources that were of benefit to the book were appropriately cited and referenced.

Johannesburg, South Africa

Michael Gbolagade Oladokun
Clinton Ohis Aigbavboa

Contents

About the Authors

Dr. Michael Gbolagade Oladokun is a Senior Research Associate in the Human Settlement and Construction Research Centre, Department of Construction Management and Quantity Surveying, Faculty of Engineering and the Built Environment, University of Johannesburg, Republic of South Africa, and also a Senior Lecturer in the Department of Building, Faculty of Environmental Studies, University of Uyo, Nigeria. Dr. Oladokun completed his Ph.D in June 2014 at the Centre of Excellence in Sustainable Building Design of the then School of the Built Environment (now School of Energy, Geoscience, Infrastructure and Society), Heriot-Watt University, UK. His Ph.D focused on issues of energy consumption and carbon emissions in housing. Before this, Dr. Oladokun had graduated with a B.Sc. (Hons.) degree in Building from Obafemi Awolowo University, Nigeria, in 2001, and completed his M.Sc. degree in Construction Management in 2008 from the University of Lagos, Nigeria. His research interests cover sustainability issues in the built environment, modelling complex systems, system dynamics applications to construction, and construction project management in general. Dr. Oladokun's research outputs are published in high-impact academic and professional journals and refereed conference proceedings. His work on energy consumption in buildings received the 2016 Emerald Literati Award for the most outstanding paper in the *International Journal of Energy Sector Management*. He is a reviewer for a number of local and international journals such as *Sustainable Cities and Society*; *Engineering, Construction and Architectural Management*; *Facilities*; and *Ecological Indicators*, among others. He has also been a member of scientific review committee of many local and international conferences.

Prof. Clinton Ohis Aigbavboa is an Associate Professor in the Department of Construction Management and Quantity Surveying, University of Johannesburg, South Africa. Before entering academia, he was involved as quantity surveyor on several infrastructural projects, both in Nigeria and South Africa. He completed his Ph.D in Engineering Management and has published several research papers in the area of housing, construction and engineering management, and research methodology for construction students. His research interest is situated in the fields of sustainable human development, with the focus on: sustainable housing regeneration (urban renewal and binformal housing), Life Cycle Assessment in the Construction Industry, remanufacturing, leadership in low-income housing, Biomimicry, post-occupancy evaluation and green job creation. He has extensive knowledge in practice, research, training and teaching. He is currently the Vice Dean: Postgraduate Studies, Research and Innovation in the Faculty of Engineering and the Built Environment and the Head: Sustainable Human Settlement and Construction Research Centre, University of Johannesburg, South Africa. He is also an author of two research books that were published with Springer Nature and CRC Press. He is currently the editor of the Journal of Construction Project Management and Innovation (accredited by the DoHET) and has received national and international recognition in his field of research. The South Africa National Research Foundation rates him as a Young researcher with the potential of establishing himself within a five-year period.

Acronyms and Abbreviations

ABM	Agent-based modelling
ANT	Actor–network theory
BBN	Bayesian belief network
BRE	Building Research Establishment
BREDEM	Building Research Establishment's Domestic Energy Model
BREHOMES	Building Research Establishment's Housing Model for Energy Studies
BSRIA	Building Services Research Information Association
CDEM	Community Domestic Energy Model
CHM	Cambridge Housing Model
CLD	Causal loop diagram
CM	Configuration modelling
CREEM	Canadian Residential Energy End-use Model
DDM	Domestic Dwelling Model
DECarb	Domestic Energy and Carbon Model
DECC	Department of Energy and Climate Change
FL	Fuzzy logic
HECCE	Household energy consumption and carbon emissions
IEA	International Energy Agency
MA	Morphological analysis
ONS	Office for National Statistics
SAP	Standard Assessment Procedure
SD	System dynamics
SFD	Stock and flow diagram
SNA	Social network analysis
STS	Sociotechnical systems

TPB	Theory of planned behaviour
UKDCM	UK Domestic Carbon Model
UNDESA	United Nations Department of Economic and Social Affairs
UNFCCC	United Nations Framework Convention on Climate Change

List of Figures

List of Tables

Chapter 1
General Introduction

1.1 Background and Significance

Governments at different levels globally are urgently seeking solutions to the problems emanating from energy consumption and carbon emissions in all spheres of the economy. This is because of the challenge of climate change and other related effects as a result of carbon emissions. For example, the evidence from the United Nations Department of Economic and Social Affairs [36] suggests that the climate change effects due to carbon emission could cause an increase in global temperature of up to 6 °C. This would invariably result in extreme weather conditions. Consequently, different initiatives and schemes of government have targeted a number of policies aimed at reducing energy and carbon emission, and the housing sector of the economy is no exception. In the UK (UK), based on the evidence from the Office for National Statistics (ONS) [28], energy consumption in buildings alone was about 42.3% of which the domestic sector accounted for around 27.5% of the total UK's energy consumption in the year 2008. Correspondingly, housing carbon emissions stand at about 26% of the total UK carbon emissions [24]. It is against this background that the housing sector of the economy has been chosen as a focal point for mitigation and adaptation agendas. As such, the UK Government, for example, has initiated a number of strategies aimed at reducing housing energy and carbon emissions (HECE). This is mainly due to the importance accorded this sector of the economy in realising a target of 80% reduction by 2050 based on the 1990 level as enshrined in the Climate Change Act of 2008.

From the foregoing, the menace posed by carbon emissions and other climate change-related effects has made it extremely difficult to accurately predict the energy and carbon emissions performance of dwellings once occupied [2, 34]. Way and Bordass [37] posit that dwellings are not only becoming more complex, but tighter energy and other environmental regulations are also increasing pressure regarding their greater predictability. Further, the outcomes of several studies have

M. G. Oladokun and C. O. Aigbavboa, *Simulation-Based Analysis of Energy and Carbon Emissions in the Housing Sector*, Green Energy and Technology,
https://doi.org/10.1007/978-3-319-75346-1_1

indicated that design predictions are not just the same as the operational outcomes of dwellings once occupied. One of the reasons advanced by the Building Services Research Information Association (BSRIA) [7] is because of the complex technology currently in use in order to allow dwellings to reach their targets of energy and carbon emissions reductions. Mahdavi and Pröglhöf [21] submitted that '*…the presence and actions of dwellings occupants have a significant impact on the energy and carbon emission performance of dwellings*'. Additionally, a number of researchers now attach much importance to occupants[1] and their behaviour in and around dwellings and advocate their inclusion while evaluating the energy and carbon emissions performance of dwellings [8, 12, 16, 26, 29, 34, 35, 38].

Therefore, it has been established that there is still much to do regarding housing energy when it comes to dwellings-occupants-environment interactions. Stevenson and Rijal [34] argue that one of the areas of uncertainty with which researchers are still struggling is in finding the means to establish a concrete methodology that links the technical aspect of dwellings' energy consumption with that of dwellings' occupants. This is with the view to capturing the effects of occupants on household energy consumption. Previous studies in this area of dwellings-occupants-environment interactions mainly focused on occupants' interactions with the control systems and devices put in place in dwellings [3–5, 10, 14, 15, 17–20, 22, 23, 25, 30, 32, 34]. The main thrust of the majority of these studies is on occupants' behaviour towards the control of windows for thermal comfort as well as lighting and shades for proper illumination in dwellings. Various models were recommended for use to predict these actions of occupants based on quantitative data collected and analysed. Generally, dwellings need to be acknowledged as being dynamic, and interactions of operators, occupants, and designers all influence the way they perform in terms of energy consumption and carbon emissions.

It is noteworthy that dwellings as a system on their own are engineered using tested components and generally reliable systems, whereas the occupants' aspect thereof can be unreliable, variable, and perhaps even behave irrationally. Borgeson and Brager [3] claim that owing to complexity in terms of energy consumption and carbon emissions, dwellings are now behaving in a nonlinear and irrational way that calls for an approach that is able to cope with this kind of complexity. Hitchcock [16] and Borgeson and Brager [3] argue that researchers are finding it difficult to predict the occupants' behavioural aspect of energy consumption in dwellings. This is mainly because the fundamental approach on which energy consumption models are based is quite different from that of the occupants' aspect. While energy models that try to capture the behaviour of occupants towards the opening of windows [3], for example, make use of a linear relationship of temperature difference, the actual actions consequently performed by the occupants follow a nonlinear and unpredictable way, which makes modelling this difficult.

[1]Occupant(s) and householder(s) are interchangeably used throughout this book to have the same meaning.

One of the breakthroughs proposed by Borgeson and Brager [3] is to model occupants' behaviour using stochastic algorithms and map this with the climate data. These models are deficient in the sense that they still face the challenge of integrating the occupants' behavioural aspect with energy models. Furthermore, the UK Government's Standard Assessment Procedure (SAP) for the energy rating of dwellings [6] that tries to assign energy rating to dwellings incorporates a number of variables into their calculations. Unfortunately, the SAP fails to capture the variables related to the individual characteristics of the household occupying the dwelling, for example, household size and occupants' behaviour, amongst others [6]. This then shows that the calculation may be deficient because of the lack of inclusion of these occupant-related variables. It is therefore necessary to explore ways of significantly improving predictability of dwellings' energy consumption and carbon emissions by demonstrating a novel approach that takes into consideration the challenge of the occupants' aspect of energy consumption in dwellings.

Undoubtedly, the integration of the dwellings' occupants' aspect with that of the dwellings' characteristics/parameters regarding energy consumption in buildings sits squarely within the sociotechnical systems (STS) approach of the systems-based methodology of scientific inquiry. As pointed out earlier, dwellings as a system are seen to comprise two subsystems: a physical subsystem that relates to the dwellings' characteristics/parameters (technical system) and a human subsystem regarding the occupants' actions within the dwelling (social system). Dwellings as a system are affected should there be any change to either the technical or social systems. Invariably, any change to the technical system will have effects on the physical subsystem; likewise, any change in the social system will have corresponding effects on the human subsystem (Fig. 1.1). On one hand, some changes to the technical system may have an indirect influence on the human subsystem, while on the other hand, some changes to the social system may have an indirect influence on the physical subsystem as well (Fig. 1.1).

It should not be forgotten that the dwellings as a system relate with the outer environment, which has both a direct and indirect influence on the technical and social systems alike. Any change in the outer environment elements will definitely influence the behaviour of these technical and social systems. This will consequently have effects on household energy consumption and associated carbon emissions. This then presents a complex kind of system that calls for an approach that is able to cope with this type of situation. However, it necessary to note that engineering models can only deal with the changes to the technical system and social models can only cope with the changes to the social system. For example, within the energy sector, modelling energy consumption and carbon emissions has been purely based on an econometric [9], statistical [11], or building physics [33] method. One of the main thrusts of this research book is to present an approach that links this phenomenon, aids in its understanding, and offers ways of testing different strategies for reducing energy consumption and carbon emissions in the housing sector. This is in order to contribute to the carbon emissions' reduction target of the UK Government. Notably, there are a number of variables at play here. These variables are interrelated and depend on one another. Among them are the variables

Fig. 1.1 Interactions between the social and technical systems

that are related to the interaction of dwellings themselves with the outer environment as well as the interaction of occupants with the systems put in place to operate dwellings in a sustainable way. All these present a complex system.

Climatic variables (outer environment element), for example, are unpredictable as any change in these (e.g. in terms of external temperature, rainfall) may haveeffects on heating, and ventilation, amongst others. They are then likely to trigger a response from the occupants to react appropriately to this situation in terms of heating and the use of hot water, for example. The reactions from occupants too still largely depend on a number of determinants (e.g. demographic, cultural, and economic variables, and behaviour). An analysis of this scenario presents a complex system that has multiple interdependencies with multi-causal relationships. The variables at play here are both 'soft' and 'hard', and their behaviour changes in a nonlinear way over time with multiple feedback loops. However, these variables are difficult to predict and keep under control. The situation described above illustrates an example of the STS problem of household energy consumption and carbon emissions. This research book then delves into the issue of STS of housing energy and carbon emissions with a view to adding to the understanding of the complex nature of energy issues of the housing sector. This is based on a novel approach proposed to policy makers that is capable of testing different strategies and interventions regarding the future profiles of energy and carbon emissions of the housing sector.

This research book then seeks to find answers to the following questions:

1. What are the social and technical variables influencing household energy consumption and carbon emissions?

2. What are the simulation approaches currently in use to forecast housing energy consumption and carbon emissions?
3. What is the most suitable simulation approach to model the complex sociotechnical systems of housing energy and carbon emissions?
4. How could the influence of these sociotechnical systems' variables on housing energy consumption and carbon emissions be modelled and predicted using a pragmatic approach?
5. What are the effects of energy efficiency measures, occupants' behavioural change, and energy prices on household energy consumption and carbon emissions?

1.2 Purpose and Objectives

There is a considerable amount of extant literature in the area of energy and carbon emissions in housing, especially materials that are focused on the energy analysis of household consumption, sustainable energy consumption and society, energy efficiency in housing, and energy performance of buildings, amongst others [1, 13, 27, 31]. This existing literature on the subject is not limited to textbooks, research books, book chapters, journal articles, and conference proceedings articles. The materials provide in-depth information regarding the areas they cover. For example, Pachauri [31] used the input–output approach to analyse household energy use in India. This research book is not a repetition of those materials. It looks at the complexity involved in the housing energy and carbon emissions from the systems thinking perspective.

The purpose of the research in this book is to describe the sociotechnical systems of housing energy consumption and carbon emissions, focusing specifically on the UK housing stock. The book describes a model capable of testing a range of possible futures regarding housing energy consumption and carbon emissions. The book is capable of improving the readers' understanding of the complex nature of energy consumption and carbon emissions in the housing sector. It attempts to evaluate the developed model by using a number of scenarios to illustrate the possible future of energy consumption and carbon emissions for the UK housing stock. These scenarios are used to explore the possibility of achieving the carbon emission reductions of about 80% by 2050 as enshrined in the Climate Change Act of 2008.

Some of the key objectives that will be addressed in the subsequent chapters include the following:

1. Identification of the sociotechnical variables influencing household energy consumption and carbon emissions.
2. Review of the modelling approaches used in forecasting housing energy consumption and carbon emissions.

3. Identification of the most suitable modelling approach to conceptualise the complex sociotechnical systems (STS) of housing energy consumption and carbon emissions.
4. Development of the dynamic model of the sociotechnical systems of household energy consumption and carbon emissions.
5. Application of the developed model to evaluate the effects of energy efficiency, occupants' behavioural change, and energy prices on household energy consumption and carbon emissions.

1.3 Organisation of the Book

The logical structure of how this research book has been arranged is shown in Fig. 1.2. Chapter 2 of the book contains a review of extant literature relating to energy consumption and carbon emissions in the housing sector. The chapter analyses issues relating to energy consumption in domestic buildings together with energy policy and carbon emissions targets, especially in the UK. Additionally, the chapter reviews the theoretical frameworks underpinning the housing energy consumption and carbon emissions. The sociotechnical variables influencing housing energy and carbon emissions are identified. The chapter also critically reviews the extant literature to reveal the epistemological issues relating to HECE models in order to critique different energy models previously or currently in use by assessing their strengths and weaknesses.

The main thrust of this research book is modelling the STS of HECE. Chapter 3 then presents an overview of the systems-based approach of scientific inquiry from where the STS theory emanates. This chapter therefore critically examines the tenets of the systems-based approach of scientific inquiry as overall umbrella under which the STS theory resides. Further to this, the STS theory is critically reviewed. This is then followed by a critical appraisal of different techniques to model the STS as identified in extant literature. The critiques of these different approaches are undertaken in a bid to identify the most suitable modelling technique to conceptualise the problem in this research book.

Chapter 4 discusses the system dynamics modelling approach as applied in this research book. The chapter gives an historical overview of the system dynamics approach. Also, the philosophical knowledge base underpinning the system dynamics approach in general is discussed in Chap. 4. This is to provide the epistemological and ontological issues of the system dynamics approach as modelling technique to conceptualise the HECE issues addressed in this research book. Moreover, issues regarding the procedures for using the system dynamics as well as the development of the model algorithms are all discussed in this chapter.

Chapter 5 applies the system dynamics approach to the UK housing sector. The chapter reports the model conceptualisation stage of the system dynamics approach. It also explains the key variables in the model, including the boundary of the model

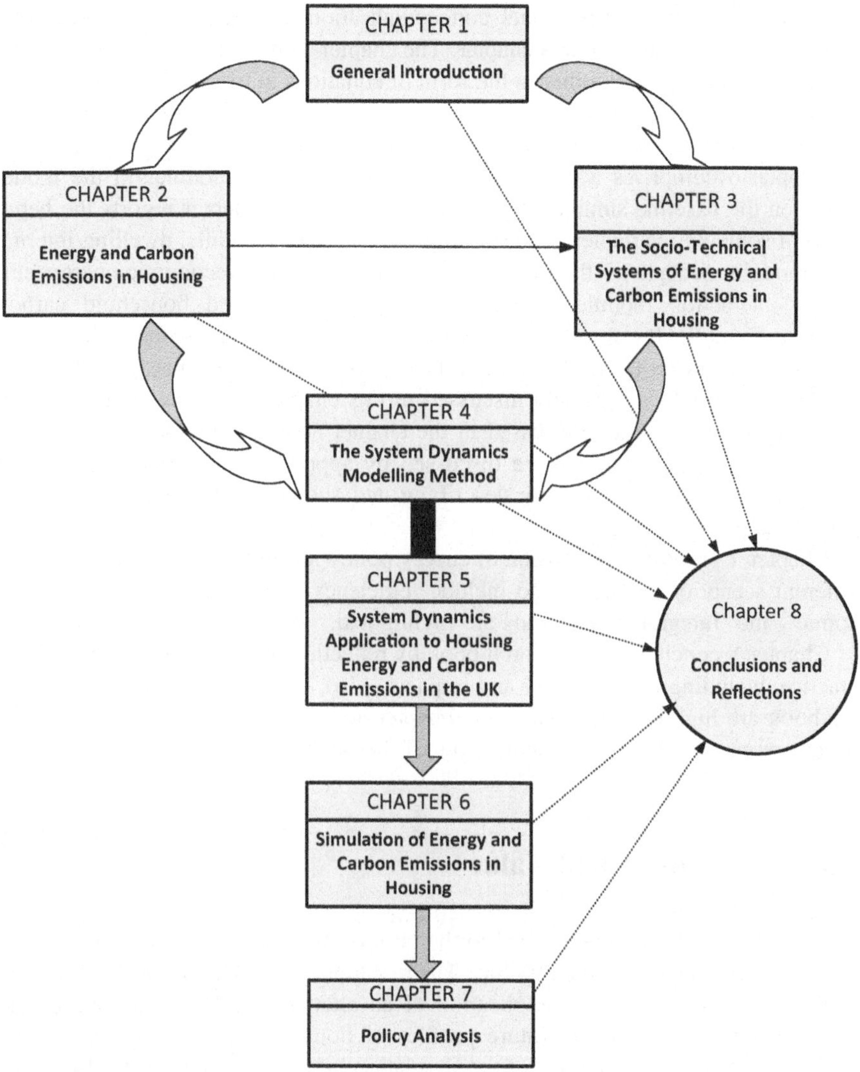

Fig. 1.2 Structure of the book

which is illustrated using a model boundary chart. Reference modes of key variables in the model are illustrated as well. Furthermore, the chapter establishes the causal relationships among the variables hypothesised to influence HECE. These are arranged into six different modules to reflect the causal loop diagram (CLD) for the population/household module, CLD for the dwelling internal heat module, CLD for the occupants' thermal comfort module, CLD for the climatic-economic-energy efficiency interaction module, CLD for the household energy module, and CLD for the household carbon emissions module. The report on the transformation of the

CLDs produced under the model conceptualisation to stock and flow diagrams (SFDs) is also presented in this chapter. The chapter shows how the variables in the model are related to each other in the form of equations in readiness for simulation. The equations developed in this chapter are for baseline simulation. The SFDs are arranged based on the six different modules of the model as explained in Chap. 5.

Chapter 6 comprises a discussion of the behaviour of variables in the model based on the baseline simulation performed. That is, the chapter reports the behaviour of some key variables in the population/household module, dwelling internal heat module, occupants' thermal comfort module, climatic-economic-energy efficiency interaction module, household energy module, and household carbon emissions module. It is noteworthy that the behaviour exhibited by the main outputs of this model in terms of the household energy and carbon emissions based on end-uses of HECE is explicitly discussed in this chapter. Issues relating to model testing and validation are discussed in the chapter as well. The model testing and validation process developed are discussed by appraising the test and validation types. This is then followed by describing and showing the results of each test performed.

Chapter 7 presents a discussion of energy policy formulations and analysis. Four different scenarios formulated to include 'efficiency', 'behavioural change', 'economic', and 'integrated' scenarios are highlighted.

Chapter 8 concludes the research book by reflecting on the value of the book for practice, including its originality and impacts. Also, the limitations of the model in the book are highlighted. Finally, the chapter addresses the recommendations for further research based on the limitations of the book.

1.4 Contribution and Value

This research book has contributed to the pool of materials available in the area of housing energy and carbon emissions. The research reported in this book is original as the originality lies in the application of an alternative simulation approach to model issues relating to the future profiles of housing energy consumption and carbon emissions. The outcome of the simulation in this book can be used by policy makers as a policy advice mechanism. The research effort within this book has brought a number of other contributions to light. The major contribution is the application of the system dynamics approach to simulate the sociotechnical issues relating to housing energy and carbon emissions which can be used for decision making over time.

The research reported in this research book indicates the first system dynamics modelling efforts applied to the entire UK housing sector in order to model the household energy consumption and carbon emissions based on the complex sociotechnical interactions of the influencing variables. Further, the developed conceptual model, which is the system thinking aspect of the model, can improve the theoretical understanding of the complex intrinsic interrelationships that exist among the sociotechnical variables influencing housing energy consumption and carbon emissions. The model developed in this research book can be used to simulate and predict the future profiles of UK household energy consumption and carbon emissions over different time horizons.

The developed model has the capability of producing a clear understanding of household energy consumption and the carbon emissions associated with it. This can serve as a decision-making policy tool with the capability of directing policy decisions by testing the effect different policy scenarios such as energy efficiency improvements and behavioural change are likely to have on household energy consumption and carbon emissions. The insights generated will allow policy makers to make informed decisions regarding any future policy formulations concerning energy and carbon emissions within the UK housing sector. The developed model is also capable of modelling and exploring the potential rate at which the carbon emissions reduction targets are being achieved within the UK housing sector.

Further to the above, the output of the research reported in this book has a number of implications for research, practice, and society in general. In the first instance, the complex intrinsic interrelationships among the STS of dwellings, occupants, and environment of housing energy consumption and carbon emissions from the systems thinking perspective can extend the frontier of knowledge in system dynamics applications to housing energy and carbon emissions. For example, the causal diagrams can lead to theory building by other interested researchers.

Furthermore, the research book has implications for practice. The developed model in this research book builds on the existing modelling efforts which are traditionally restricted to building physics and regression-based forecasting. The simulation in this research book generates a number of new insights into the future using a nondeterministic systems approach. The main value of the developed model in this research book is that it is highly transparent as all the variables and algorithms in the model can be scrutinised. Therefore, the model in the book will benefit the software developers who can prototype it into other suitable user-friendly platforms. Finally, the outcome of this research has implications for society by providing the policy makers with a decision-making tool upon which different scenarios regarding HECE can be tested before implementation.

1.5 Summary

This chapter introduced this research book by describing and explaining the background issues surrounding the modelling and simulation of energy consumption and carbon emissions, especially in the housing sector of the economy. The chapter explained the main purpose and objectives of this research book. The purpose of the book is primarily to describe the use of the system dynamics method to model the housing energy and carbon emissions by focusing specifically on the UK housing stock. This chapter also provided the structure of this research book, as well as its contributions and value.

References

1. Boemi S, Irulegi O, Santamouris M (2015) Energy performance of buildings: energy efficiency and built environment in temperate climates. Springer, Berlin
2. Bordass W, Cohen R, Field J (2004) Energy performance in non-domestic buildings: closing the credibility gap. Paper presented at the building performance congress 2004, Frankfurt, Germany
3. Borgeson S, Brager G (2008) Occupants' control of windows: accounting for human behaviour in building simulation. Centre for the Built Environment, University of California —Internal Report, October 2008
4. Bourgeois D, Reinhart C, Macdonald IA (2005) Assessing the total energy impact of occupant behavioural response to manual and automated lighting systems. In: Proceedings of the IBPSA international conference, Montreal, Canada, pp 99–106
5. Bourgeois D, Reinhart C, Macdonald I (2006) Adding advanced behavioural models in whole building energy simulation: a study on the total energy impact of manual and automated lighting control. Energy Build 38:814–823
6. Building Research Establishment (2011) The government's standard assessment procedure for energy rating of dwellings. 2009 edition incorporating RdSAP 2009, Watford, UK
7. Building Services Research Information Association (BSRIA) (2011) Introducing soft landings. Available at: http://www.bsria.co.uk. Accessed 18 July 2011
8. Dietz T, Gardner GT, Gilligan J, Stern PC, Vandenbergh MP (2009) Household actions can provide a behavioural wedge to rapidly reduce US carbon emissions. Sustain Sci 106 (44):18452–18456
9. FitzGerald J, Hore J, Kearney I (2002) A model for forecasting energy demand and greenhouse gas emissions in Ireland. Working paper no. 146, The Economic and Social Research Institute, Dublin
10. Fritsch R, Kohler A, Nygard-Ferguson M, Scartezzini JL (1990) A stochastic model of user behaviour regarding ventilation. Build Environ 25(2):173–181
11. Fung AS (2003) Modelling of national and regional energy consumption and associated greenhouse gas emissions. Ph.D. thesis, Dalhousie University
12. Gill ZM, Tierney MJ, Pegg IM, Allan N (2010) Low-energy dwellings: the contribution of behaviours to actual performance. Build Res Inf 38(5):491–508
13. Goldblatt DL (2005) Sustainable energy consumption and society: personal, technological, or social change? Springer, Berlin
14. Haldi F, Robinson D (2009) Interactions with window openings by office occupants. Build Environ 44(12):2378–2395. https://doi.org/10.1016/j.buildenv.2009.03.025
15. Herkel S, Knapp U, Pfafferott J (2005) A preliminary model of user behaviour regarding the manual control of windows in office buildings. International Building Performance Simulation Association (IBPSA)
16. Hitchcock G (1993) An integrated framework for energy use and behaviour in the domestic sector. Energy Build 20:151–157
17. Humphreys MA, Nicol JF (1998) Understanding the adaptive approach to thermal comfort. ASHRAE Trans 104(1):991–1004
18. Humphreys MA, Nicol JF, Tuohy P (2008) Modelling window-opening and the use of other building controls. In: Proceedings of the AIVC conference, Tokyo, Japan

19. Hunt D (1979) The use of artificial lighting in relation to daylight levels and occupancy. Build Environ 14:21–33
20. Kabir E, Mohammadi A, Mahdavi A, Pröglhöf C (2007) How do people interact with buildings environmental systems? Build Simul, 689–695
21. Mahdavi A, Pröglhöf C (2009) User behaviour and energy performance in buildings. In: 6. Internationale Energiewirtschaftstagung, Vienna University of Technology, Vienna
22. Mahdavi A, Kabir E, Mohammadi A, Lambeva L, Pröglhöf C (2006) User interactions with environmental control systems in buildings. In: Proceedings of the PLEA 2006—23rd international conference on passive and low energy architecture, Geneva, Switzerland, 6–8 Sept 2006
23. McDermott H, Haslam R, Gibb A (2010) Occupant interactions with self-closing fire doors in private dwellings. Saf Sci 48(10):1345–1350. https://doi.org/10.1016/j.ssci.2010.05.007
24. Natarajan S, Padget J, Elliott L (2011) Modelling UK domestic energy and carbon emissions: an agent-based approach. Energy Build 43:2602–2612
25. Newsham GR (1994) Manual control of window blinds and electric lighting: implications for comfort and energy consumption. Indoor Environ 3:135–144
26. Nicol F, Roaf S (2005) Post-occupancy evaluation and field studies of thermal comfort. Build Res Inform 33(4):338–346
27. Nieboer N, Tsenkova S, Gruis V, Van Hal A (2012) Energy efficiency in housing management: policies and practice in eleven countries. Routledge, Abingdon-on-Thames
28. Office for National Statistics (2009) Digest of UK energy statistics 2009. The Stationery Office, London
29. Okhovat H, Amirkhani A, Pourjafar MR (2009) Investigating the psychological effects of sustainable buildings on human life. J Sustain Dev 2(3):57–63
30. Parys W, Saelens D, Hens H (2010) The influence of stochastic modelling of window actions on simulated summer comfort in office buildings. In: Proceedings of the AIVC 2010 conference on low energy and sustainable ventilation technologies for green buildings, Seoul, 26–28 Nov 2010, pp 1–20
31. Pachauri S (2007) An energy analysis of household consumption: changing patterns of direct and indirect use in India. Springer, New Delhi
32. Rijal H, Tuohy P, Humphreys M, Nicol J, Samuel A (2011) An algorithm to represent occupant use of windows and fans including situation-specific motivations and constraints. Build Simul 4(2):117–134
33. Shorrock LD, Dunster JE (1997) The physically-based model BREHOMES and its use in deriving scenarios for the energy use and carbon dioxide emissions of the UK housing stock. Energy Policy 25(12):1027–1037
34. Soldaat, K (2006) Interaction between occupants and sustainable building techniques. In ENHR Conference: Housing in an expanding Europe: theory, policy, participation and implementation, Ljubljana, Slovenia, 2–5 July, pp 1–15.
35. Stevenson F, Rijal HB (2010) Developing occupancy feedback from a prototype to improve housing production. Build Res Inf 38(5):549–563
36. United Nations Department of Economic and Social Affairs (2010) Buildings and construction as tools for promoting more sustainable patterns of consumption and production. Sustain Dev Innov Briefs 9. Available at: http://www.un.org/esa/sustdev/publications/innovationbriefs/index.htm. Accessed 23 May 2011
37. Way M, Bordass B (2005) Making feedback and post-occupancy evaluation routine 2: soft landings—involving design and building teams in improving performance. Build Res Inf 33 (4):353–360
38. Yun GY, Steemers K (2011) Behavioural, physical and socio-economic factors in household cooling energy consumption. Appl Energy 88:2191–2200

Chapter 2
Energy and Carbon Emissions in Housing

2.1 Introduction

Within the research community, there is a general consensus that the threat of global warming as a result of climate change will increase. As a result, different strategies have evolved targeting carbon emissions reductions. The housing sector is therefore at the centre of these reduction targets. This chapter discusses issues of energy consumption and carbon emissions in the housing sector together with how energy policy has evolved over the years. Also, the chapter reviews previous studies that serve as the theoretical framework underpinning models for the energy and carbon emissions in housing. This forms the basis for reviewing the sociotechnical variables influencing energy consumption and carbon emissions in housing.

The chapter also identifies the arrays of energy models that have evolved over the years together with their capability of analysing energy consumption and its associated carbon emission trends in the housing sector of the economy. This is as a result of the growing need to curtail the carbon emissions globally in order to achieve the climate stabilisation goals based on the climate change threat as mentioned above. For example, in the UK, the need to realise the target of carbon emissions as stipulated in the Climate Change Act of 2008 has rapidly spurred the emergence of many energy models, especially in the domestic sector of the economy, to analyse different strategies and schemes of the government. These kinds of models are therefore required in order to guide an understanding of the effects of different strategies and schemes of the government before they are implemented. The chapter concludes by critiquing the energy and carbon emission models in the housing sector.

M. G. Oladokun and C. O. Aigbavboa, *Simulation-Based Analysis of Energy and Carbon Emissions in the Housing Sector*, Green Energy and Technology,
https://doi.org/10.1007/978-3-319-75346-1_2

2.2 Energy Consumption in Housing

There is a general consensus within the research community regarding the threat of global warming as a result of climate change, namely that it is due to an increase in greenhouse gases emitted into the atmosphere because of the profligate use of fossil fuels [40, 53]. The greenhouse gases mainly include carbon dioxide (CO_2), methane (CH_4), hydrofluorocarbons (HFCs), nitrous oxide (N_2O), perfluorocarbons (PFCs), and sulphur hexafluoride (SF_6). The contribution of each of these emissions to climate change in the form of global warming varies considerably. For example, carbon dioxide emission is adjudged to be the most worrying of these gases as its levels in the atmosphere are rising so rapidly [106]. To date, the concentration of carbon dioxide emissions in the atmosphere has increased profoundly from approximately 280 parts per million (ppm) in 1750 to about 380 ppm in 2005 with the potential of reaching about 540–970 ppm by the end of the twenty-first century [53]. The IPCC [53] lists the effects of these increases in carbon dioxide emissions on the global climate to include the following:

- An increase in the globally averaged surface temperature of 1.4–5.8 °C;
- An increase in the global mean sea level of 9–88 cm;
- A decrease in snow cover and the extent of sea-ice in the Northern Hemisphere;
- Changes in weather patterns which are likely to result in an increase in globally averaged precipitation and the occurrence of extreme weather events;
- The possibility of famines and population migrations; and
- The extinction of rare species and the loss of habitats.

The contributing factor to these emissions is not different from the assertion of Harris [40] suggesting the profligate use of fossil fuels. However, burning forest is also a significant contributor, according to Terry [106]. Importantly, Terry [106] posits that emissions from both methane and nitrous dioxide are the most potent emissions when compared to all other greenhouse gases. However, they are in trace amounts in the atmosphere compared to the high concentration of carbon dioxide, suggesting they have less of an overall impact. This then explains the reason why the term ‘carbon emission’ is always used to mean greenhouse gases.

Obviously, if the concentration of carbon emissions is allowed to continue to grow unabated, it will undoubtedly have substantial repercussions, both politically and socio-economically [61]. Therefore, a general consensus was reached within the world’s governments to significantly reduce what the carbon emissions will be in this twenty-first century. This is reflected in the Rio Summit of 1992 where the United Nations Framework Convention on Climate Change (UNFCCC) was signed, committing developed nations to significantly reduce their carbon emissions profiles [63]. Subsequently, there was series of follow-ups regarding the UNFCCC agreement of 1992 which was ratified in 1993, for example, the World Climate Conference of 1997 in Kyoto, Japan and the Copenhagen Summit in 2009. The latest of these agreements was the declaration of the 2015 United Nations Climate

Change Conference in Paris. At the Kyoto conference, for instance, a legally binding agreement was reached mainly to remove the emissions from the six aforementioned causes of climate change. Among the developed countries committed to significantly reducing their carbon emissions' profile is the UK. As such, the UK has since then followed a path aimed at reducing its carbon footprints. In this regard, the housing sector in the UK contributes substantially to the UK's total carbon emissions, which in this case is about 26% of the UK's total emissions [85]. Since then reductions in energy consumption within the housing sector have been a target.

Focusing on the housing sector, Harris [40] argues that approximately 50% of energy use and carbon emissions are as a result of energy used for heating (space and water heating), cooling, and lighting in buildings. Similarly, Urge-Vorsatz et al. [111] and the International Energy Agency (IEA) [52] point out that buildings worldwide account for 122 EJ of final energy in 2010, which translates to 33% of the total final energy and 54% of electricity. Accordingly, this amount of final energy corresponds to about 9 Gt of carbon emissions [60, 110].

The figure of energy consumption in buildings, basically for space and water heating, cooling, and lighting, has been estimated to be around 30–60% of primary energy, especially in Western Europe with about half of this amount used in housing within the UK [40]. The end-uses (i.e. space and water heating, cooking, lighting, and appliances) of delivered energy in the UK housing sector suggest that they vary considerably within different dwelling types. Harris [40] and Terry [106] argue that about 58% of delivered energy is used for space heating and when combined with water heating, it rises to about 82% within the UK housing sector. This suggests that space and water heating have the potential of shaping household energy consumption. As such, a carbon emissions reductions policy is required to target these end-uses.

2.3 Energy Policy and Emissions Targets

Over the years, the UK energy sector has witnessed tremendous improvements and changes in energy policy. Principally, energy policy has been shaped by two major factors: firstly, as a result of market liberalisation of 1980s, which saw the state-controlled energy companies privatised and the Department of Energy dismantled; and secondly, the rising threats of climate change effects as highlighted by the Rio Summit of 1992 [63] have also significantly shaped energy policy within the UK. This single factor has risen up the agenda in the UK; a signatory to the Rio Summit of 1992 as well as other agreements, especially the 2015 Paris agreement, the UK is committed to the reduction of carbon emissions. As a result of this, the UK Government published its white paper on energy entitled Our Energy Future—creating a Low-Carbon Economy in 2003. In this white paper, the UK Government is committed to a 60% reduction in carbon emissions by the year 2050 [25].

Undoubtedly, the world of energy is changing rapidly, and in view of such a constraint, a review of the energy policy is inevitable. Policy frameworks such as the Energy Review Report of 2006, the Energy White Paper of 2007, the Climate Change Act of 2008, the UK Low-Carbon Transition Plan of 2009, and the Energy Bill of 2012–2013 all tend to shape energy policy, mainly to stem the rising tide of carbon emissions and to effect drastic reductions.

Consequently, the Climate Change Act provides a legally binding pathway towards carbon emission reductions by reducing 22% of carbon emissions between 2008 and 2012 relative to the base year 1990. In addition, the Act stipulates that reductions of 28% are to be achieved between 2013 and 2017, while a 34% reduction is required between 2018 and 2022. The extent of these reductions is currently uncertain. Additionally, the Act specifies that carbon emission reductions of 50% are envisaged for between the years 2023 and 2027, while 80% is to be achieved by 2050 relative to the base year 1990. These savings are to be achieved within all spheres of the economy.

Different studies have, however, highlighted the potential of the housing sector to contribute significantly to these reductions [11, 27, 72, 76]. As such, some of these studies have indicated the areas of possible policy targets. For example, Levine et al. [72] highlight the importance of technological developments and cultural and behavioural choices as possible areas of policy formulation. In the UK, however, a number of policy targets are in place within the housing sector. These policies target both new and existing homes. For example, Ko and Fenner [68] point out that there is a range of policy frameworks on energy efficiency for new homes in the UK as shown in Table 2.1. The frameworks are as a result of various energy policy reviews as previously highlighted. Table 2.1 shows the programmes that have been firmed up and have the capability of improving the energy efficiency profile of new-built homes in the UK.

Having discussed the issues of energy policy in housing, it is equally important to review the trends regarding the empirical studies relating to energy and carbon emissions in housing. As such, the next section of the book discusses the trends in housing energy and carbon emissions' (HECE) research.

2.4 Trends in Housing Energy and Carbon Emissions' Research

The preceding section laid the foundation regarding the issues relating to energy and carbon emissions in the housing sector generally. In this section, previous empirical studies on the sociotechnical systems of household energy consumption and carbon emissions are reviewed.

Table 2.1 UK government policy framework on energy efficiency in new homes

Policies	Plans	Programmes
• Energy Bill, 2012–2013 • UK Low-Carbon Transition Plan, 2009 • Climate Change Bill 2008 (draft 13 March 2007) • Energy White Paper 2007 • Pre-Budget Report 2006 • Climate Change and Sustainable • Energy Act 2006 • EU Energy Performance of Buildings Directive 2002 • Housing Act 2004 • Electricity Act 1989 • Gas Act 1986	• Building a greener future: towards zero carbon • Development (December 2006 consultation paper) • Climate change programme (revised in 2006) • Energy efficiency: the government's plan for action 2004 • Sustainable communities plan 2003	• Building regulations, Part L1A 2006 • Code for sustainable homes • Government funding for social housing and developers only if they meet CSH level 3 or better. New houses by English partnerships to comply with CSH level 3 or better • Energy Efficiency Commitment 2 (2005–2008), succeeded by carbon emissions reduction • Target Energy Efficiency Commitment (2008–2011) for electricity and gas suppliers (usually relates to energy efficiency in existing houses) • Energy performance certificates and housing information packs • Improved metering and billing information for homeowner. In 2008–2010, free real-time electricity displays for homeowners who request one • Energy-saving trust product endorsement (energy labels) and building design information low-carbon buildings programme (funding for energy supply technologies but has energy efficiency requirements) • Stamp duty land tax exemption for zero-carbon homes • Reduced VAT rate of 5% for energy-saving materials like insulation, draught stripping, hot water, and central heating controls • Research and dialogue programmes including carbon vision programme (buildings) and Foresight (sustainable energy management and the built environment)

Adapted with some extension from Ko and Fenner [68]

2.4.1 The Theoretical Framework Underpinning the Energy and Carbon Emissions in Housing

Within the housing energy studies, there is a plethora of frameworks underpinning the theoretical knowledge base for conceptualising housing energy and carbon emissions. These have contributed significantly regarding the tools used in analysing and formulating the housing energy and carbon emissions' policy. Keirstead [66] advocates two main domains for those theoretical frameworks, namely the disciplinary and integrated domains. Keirstead [66] submits that the disciplinary framework, for example, the engineering or economic domain, has been the main framework guiding and dominating the energy policy over the years. Keirstead [66] further argues that relying on either the engineering or economic models alone may not effectively capture the complex problems of energy and carbon emissions within the housing sector. Therefore, the disciplinary approach alone may be limited.

Accordingly, the work of Natarajan et al. [85] also highlights the limitations of purely disciplinary methods of conceptualising the housing energy and carbon emissions. In their submission, they advanced that the limitations of a purely disciplinary method explain the gaps being witnessed between the actual and predicted energy consumption and carbon emissions in housing. To improve on the limitations of the disciplinary approach to energy policy conceptualisation, a number of sources of extant literature have advocated an integrated framework that encourages a multi/cross-disciplinary approach. This interdisciplinary approach has a way of capturing the interrelationships among the scientific, technological, societal, economic, and cultural, among others, of the housing energy and carbon emissions. The disciplinary and integrated frameworks for housing energy and carbon emissions are further discussed as follows:

Disciplinary framework

Over the years, studies relating to the housing energy and carbon emissions have been championed principally by four major disciplines, with each discipline illustrating its own approach/framework for solving HECE problems. These disciplines are engineering, economics, psychology, and sociology and anthropology. The engineering framework, for example, mainly illustrates the technology of HECE by estimating the HECE based on the physical laws with little or no attention to economic, sociological, or even behavioural aspects of HECE. This shows the limitation in this type of framework for its inability to capture a web of interactions among different disciplines. For example, the studies of Anderson and Clark [6] illustrate a framework for energy consumption of heating based on a heat transfer method; Stokes et al. [99] give the framework for domestic energy demand; and Hart and De Dear [41] provide the framework for weather sensitivities regarding household appliances use, among other studies. The point here is that behavioural responses to technical improvements of HECE [66], for example, are quite beyond the ambit of any purely engineering framework and this may then portend that such an engineering framework might be inadequate.

Further, the economic framework as one of the disciplinary frameworks conceptualises HECE when it comes to understanding HECE owing to the effects of income levels, energy prices and taxes [12, 37, 54, 55, 92]. As a social science-based framework, it introduces some behavioural aspects. Interestingly, Wheelock and Oughton [113] argue that, based on the available evidence, the concept depicted by the economic approach is not complete in aiding the understanding of HECE. As such, Lutzenhiser and Hackett [75] submit that the combination of the approaches as provided by both the engineering and economic theories forms the physical–technical–economic framework of HECE which has helped immensely in shaping energy policies around the globe. This feat was vehemently criticised for its inability to properly account for the human behavioural aspect of housing energy in the framework.

Based on that background, the studies in the area of psychology took up the challenge of contributing to the understanding of energy consumption behaviour. Notable among the behavioural studies is the theory of planned behaviours (TPB) of Ajzen [5]. This has contributed significantly to the behavioural aspect of HECE by serving as a theoretical knowledge base for many studies. However, the TPB framework cannot be used as a standalone framework for explaining HECE because the theory only used personal constructs such as attitudes and beliefs without any recourse to other aspects such as social-cultural-technical contexts. This then led to studies in the field of sociology and anthropology in a bid to conceptualize energy and society.

Reflecting on these approaches, it is evident that they are unlikely to capture the kind of complex problemsplaguing the energy sector, and hence, the need for a more robust approach that is capable of integrating anumber of disciplinary approaches.

Integrated framework

The argument from the foregoing reinforces the need for a more robust interdisciplinary framework to conceptualise the HECE. This then led to a combination of different disciplines that conceptualise the issue of HECE in order to aid a better understanding of energy issues and proffer adequate solutions. In this regard, a number of 'integrated' frameworks have therefore been used to aid the understanding of the HECE issues. Among those studies is the work of Van Raaij and Verhallan [112], which provides a novel approach to conceptualising energy behaviour. Their framework made use of both the physical parameters of dwellings and behavioural characteristics of households. While this work has been continually cited by many studies in the area of consumer behaviour and economic psychology, the framework is yet to be fully developed into a simulation model by both the researchers and industry practitioners. Further to the work of Van Raaij and Verhallan [112], the research of Lutzenhiser [74] proposes a cultural framework of HECE by conducting a survey of existing approaches in the fields of engineering, economics, psychology, sociology, and anthropology. The framework highlights how the dwelling occupants ('consumers') make some decisions regarding their choices that are 'culturally sensible' and 'collectively sanctioned' containing engineering and economic aspects as sub-systems in the framework. However, the framework remains a theoretical framework without any further work being done to turn the idea into simulation models.

Another study by Hitchcock [44] uses the systems theory to provide an integrated framework of energy use and behaviour in dwellings. He argues that the energy consumption patterns in dwellings need to be fully understood from the systems perspective because of the complexity involved in integrating both the technical and social phenomenon. He further contends that while the engineering models used in capturing the physical processes of dwellings and their effects on energy consumption do give a better understanding of the physical characteristics of dwellings, they fail, however, to capture the effects of the human aspect on dwellings. Additionally, he contends that the social models are used in capturing the human aspect effects and as such can influence energy consumption in dwellings. The study used the concept of sociotechnical systems to conceptualise HECE and came up with a framework. Yet, no modelling technique was proposed to capture these sociotechnical systems. To fully understand the different variables that constitute either the disciplinary or integrated framework, it is important to review the extant previous studies on the sociotechnical variables shaping the housing energy sector. As such, the following subsection reviews the literature along these lines.

2.4.2 Empirical Studies on the Sociotechnical Variables Influencing Household Energy Consumption and Carbon Emissions

The work of Hitchcock [44] sees a household as a system being defined by both a physical and a social sense. This implies that the physical household is in the form of materials and devices, whereas the social household is in the form of occupants living within the dwelling. Further, the combination and the way these two systems interact determine the energy consumption patterns of those dwellings [44]. There is therefore a third system that plays a very important role in the relationship, namely the household environment. It is universally accepted that domestic dwellings are basically to provide a comfortable environment for human activities by providing space heating, lighting, and hot water, among others. Hitchcock [44] contends that the amount of energy consumed in dwellings depends on the level of service required and the efficiency with which the dwelling can provide such a service. Correspondingly, energy consumption is driven by the needs or behaviour of occupants or by the physical characteristics of dwelling, or by both [44].

Within this clime, there are various empirical studies that have explored the sociotechnical interactions that influence HECE such as those of Hitchcock [44], Moll et al. [81], Bartiaux and Gram-Hanssen [14], Bin and Dowlatabadi [15], Yun and Steemers [118], Abrahamse and Steg [1], Kelly [67], the CIBSE [22], Tweed et al. [109], and Gram-Hanssen [36], among others. These studies generally cover the identification of affecting variables, ranking these variables based on importance, and defining the causal effects of variables on the HECE.

For example, the study of Hitchcock [44] identifies three elements within the household system, namely physical, human, and environmental components. Within the physical elements, Hitchcock [44] posits that this subsystem consists of physical parameters and variables that influence energy consumption such as the physical characteristics of dwellings in the form of their size, materials, heating system, and stock of appliances, among others; and physical variables in the form of the dwelling's internal temperature, ventilation rates, amount of hot water, and appliances use, among others. Additionally, Hitchcock [44] argues that the human subsystem consists of variables relating to the biophysical, demographic, and psychological aspects. For example, the biophysical aspect consists of variables such as occupants' thermal comfort in the form of metabolic rate, respiration, clothing. He refers to demographic variables as household income, social status, the number of occupants, and the like; whereas psychological variables relate to the individual beliefs, attitudes, knowledge, and personalities. Furthermore, Hitchcock [44] highlights that the environmental aspect of a household consists of three major elements, namely the climate system, the economic system, and the cultural system. For example, the climate system consists of external temperatures, insulation, and wind levels. Interaction of these affects the demand for heating and lighting. The economic system involves variables comprising energy prices, energy tax, and the like; whereas the cultural system embraces the general beliefs held by society, consumption habits, and the like. His studies posit that all these variables work together seamlessly to influence household energy consumption.

Yun and Steemers [118] identify six categories of variables that influence energy consumption in dwellings. These are variables that are related to climate (cooling degree days), building (total floor area, number of windows, year of construction, and type of housing unit), occupants (number of household members, total annual income, and age of householder), equipment (type of air conditioning equipment), behaviour (number of cooled rooms and frequency of air conditioning equipment use), and energy (total energy for space cooling). The study carried out a path analysis to identify the significant direct or indirect effect on cooling energy use and revealed that the climate is the most significant variable influencing cooling energy use.

In yet another study, Moll et al. [81] posit that the household energy consumption is strongly related to the sociodemographic variables, such as income and household size. They argue that households with higher incomes or with larger sizes tend to consume more energy. Moll et al. [81] use the hybrid energy analysis of household consumption as the methodological approach based on the concept of household metabolism. This approach involves some statistical analyses. The study of Bartiaux and Gram-Hanssen [14] considers some sociopolitical variables influencing household electricity consumption by comparing Denmark and Belgium. The work of Abrahamse and Steg [1] highlights psychological and sociodemographic variables influencing HECE. The research of Bin and Dowlatabadi [15] illustrates a series of variables influencing HECE in the US. Other studies by Gatersleben et al. [33] and Poortinga et al. [91] investigated some attitudinal and sociodemographic variables and found that household income and size are better explanatory variables of HECE, while environmental attitudes are weaker predictors.

There are many studies that have empirically explored the importance of occupants' behaviour regarding HECE. For example, the research of Barr and Gilg [13]: 907) examined '...the ways in which environmental action is constructed in everyday life and related to everyday practices...' and '...the extent to which there are identifiable groups of individuals with different behavioural properties that exemplify alternative environmental lifestyles and consequently from lifestyle groups'. The study used the sociopsychological approach to investigate the problem via a questionnaire survey. This study identified four clusters of individuals as 'committed environmentalists', 'mainstream environmentalists', 'occasional environmentalists', and 'non-environmentalists'. The study also investigated some variables relating to the social and environmental values of group of individuals involved in the research. Variables included in the social value are 'altruistic', 'openness to change', 'conservative', and 'egoism'. Also environmental value variables included in the research are 'faith in growth: anthropocentrism', 'spaceship earth: biospherism', and 'ecocentrism-technocentrism'. Further to these variables, the research sought to know the environmental attitudes of the respondents and included the following factors by framing the questions posed in a pro-environmental direction way: 'concern and commitment', 'moral motives', 'outcome beliefs', 'price', 'satisfactions', 'logistics', 'green consumer attitudes', 'comfort', 'environmental rights', 'awareness of norms', 'trust and responsibility', 'extrinsic motivation', 'personal instinct, 'brand loyalty', and 'personal threat'.

Isaacs et al. [56] carried out surveys on occupants' behaviour in New Zealand. This study, according to Stevenson and Leaman [100], was adjudged to be the largest survey in housing. The results of the study revealed that the occupants of the studied area are more comfortable living with a very low temperature. This is highly puzzling and surprising! Further—probing by this study to know the cause(s) of this behaviour indicates that culture, coupled with a lack of heating appliances, is responsible for this behaviour. That is, the lack of central heating may suggest that the rooms are heated on a one-by-one basis, thereby enabling the occupants to set the temperatures of each room according to their expectations. This then suggests that a room-by-room monitoring of comfort level provision will then be worth researching.

Of interest to this book is the work of Gill et al. [34] entitled Low-Energy Dwellings: The Contribution of Behaviours to Actual Performance. The theoretical knowledge base underpinning their study is the work of Ajzen [4] on the theory of planned behaviour. The work of Gill et al. [34] gives a simple statistical computation on how to actually estimate the contribution of occupants' behaviour to variations witnessed on the dwelling performance when the performance of heat, electricity, and water consumption were conducted. The study used some high-performing dwellings as their case study and a detailed post-occupancy evaluation was undertaken to reveal energy and water consumption performance, and the comfort and satisfaction of the occupants. The results of the study indicate that resource-conscious behaviours account for about 51, 37, and 11% of the variance in heat, electricity, and water consumption, respectively, for the dwellings studied. The study then shows the importance and significance that has been accorded to the human factor as the study demonstrates how the behaviour of occupants can influence the use of energy. Premised on the above, Stevenson and Leaman [100] then argue that designers should not use the behaviour of occupants as an excuse for the lack of performance of dwellings. They contend that designers need to understand the influence of occupants' behaviour on the dwellings' performance and incorporate the same appropriately in their designs.

The study of Williamson et al. [114] is another thought-provoking study that challenged the failure of regulatory provisions to capture behaviours of occupants in reality. The study investigates five award-winning dwellings in Australia for their ability/inability to meet relevant regulatory standards. The results of the study reveal that the regulatory concept of 'meeting generic needs' fails to account for the diversity of sociocultural understandings, the inhabitants' expectations and their behaviours. Consequently, however, comfort levels and low-energy consumption were unable to be predicted by the so-called standards and regulations. The study then suggests that occupants' behaviours and goals need to be captured by the standards and regulations.

Reflecting on the above reviews, it has been established that there is still more to do concerning the relationship between environment and behaviour in and around dwellings. Stevenson and Rijal [101] argue that researchers are still wrestling with this relationship and finding means to establish a concrete methodology that links the technical assessment of dwellings with that of occupants. This debate therefore sits squarely within the sociotechnical approach which recognises the fact that technological development in the built environment is influenced by human aspects. This problem can then be viewed as a sociotechnical problem involving modelling or simulation of housing energy and carbon emissions which is the main crux of this book.

2.4.3 Energy Consumption Behaviours

Different studies [45, 117] have established that energy consumption can change significantly under different consumption behaviours. These studies highlight the importance of accounting for different occupants' behaviour for it will aid in reliable and accurate estimations of dwellings' energy [10]. Many studies have established the classification of different energy behaviours. For example, a survey conducted by the Scottish Environmental Attitudes and Behaviour (SEAB) [94] classifies the environmental behaviour into five, namely as deep greens, light greens, shallow greens, distanced, and disengaged. In another study, the research of Accenture [2] classifies energy consumers in various countries around the world into eight different categories. Additionally, the Low-Carbon Community Challenge [24] classifies energy consumption behaviour into four as active energy savers, energy aware, energy ambivalent, and energy wasters. The classification of the Energy Systems Research Unit (ESRU) [28] and that of Azar and Menassa [10] is similar. For example, the ESRU [28] classifies energy consumption behaviour into profligate, standard, and frugal. Azar and Menassa [10] classify this as high-energy consumers, medium-energy consumers, and low-energy consumers to mean profligate, standard, and frugal consumption behaviour, respectively. Profligate energy consumers here refer to those occupants who over-consume energy. The standard energy consumers are those occupants who make little effort at reducing energy consumption, whereas frugal energy consumers use energy efficiently and do everything possible to reduce their energy consumption. Within this book, the energy consumption behaviour follows the classifications of both the ESRU [28] and Azar and Menassa [10].

Having reviewed the extant literature relating to the sociotechnical variables influencing the housing energy and carbon emissions, it is appropriate to now delve into the methods used in modelling housing energy and carbon emissions. As such, the next section discusses the approaches used in modelling or simulating housing energy and carbon emissions.

2.5 Methods of Modelling Energy and Carbon Emissions in Housing

This section provides a review of the underlying approaches to modelling issues relating to housing energy and carbon emissions. It is important to note that there is considerable extant literature on modelling techniques and methodologies and techniques for housing energy and carbon emissions. Among those studies are the works of Böhringer and Rutherford [16], Strachan and Kannan [102], Tuladhar et al. [108], Swan and Ugursal [104], and Kavgic et al. [64]. However, it is necessary to update these studies in line with the objectives of this book. As such, the following subsections extend the reviews conducted by the above studies.

2.5.1 Epistemology of Modelling Approaches

Over the last few decades, many studies have evolved regarding the techniques used in modelling housing energy consumption, especially at the national level. The forecasting or modelling methodologies for housing energy consumption have been argued to motley vary considerably based on the requirements, assumptions, and the models' predictive abilities [61, 64]. As such, the modelling approaches within this research circle have been overwhelmingly agreed to fall basically within two epistemic approaches, namely the top-down and bottom-up approaches [51]. However, both Kelly [67] and Kavgic et al. [64] acknowledge the recent advances regarding the epistemology of modelling approaches for housing energy consumption in terms of the development of another modelling paradigm derived from both the top-down and bottom-up approaches. This development has led to some cases where a hybrid of the two approaches has been made for more robust models to be developed based on the suggestion of Böhringer and Rutherford [16]. The main epistemic approaches to modelling housing energy and carbon emissions as drawn from the work of IEA [51] regarding both the top-down and bottom-up techniques are illustrated in Fig. 2.1.

As can be drawn/deduced from Fig. 2.1, the epistemic approach to the top-down technique differs from the bottom-up technique based on the fact that the top-down method starts with the aggregate data, which are then disaggregated down in order to have a comprehensive model. Accordingly, Johnston [61] argues that the top-down method gives a comprehensive approach to modelling and as such possesses the ability to support a high level of government's policy and decision

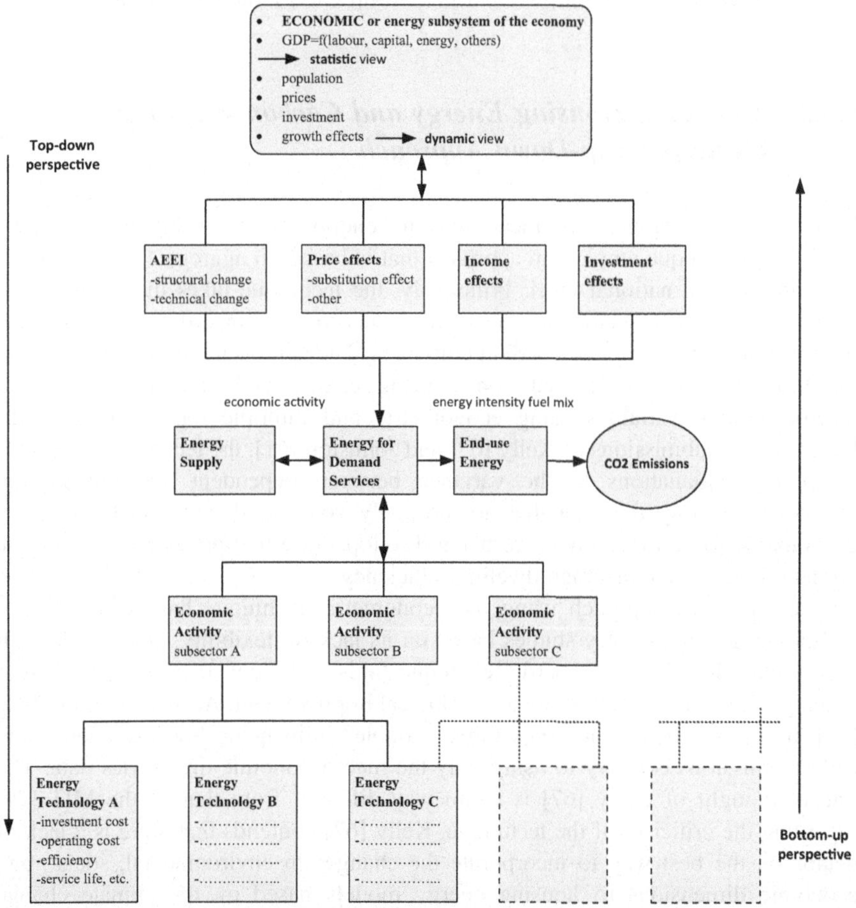

Fig. 2.1 Top-down and bottom-up modelling approaches. Adapted from IEA [51]

making. However, the bottom-up modelling technique uses the highly disaggregated data and ends up aggregating them as far up as possible. Based on the submissions of Johnston [61] and the IEA [51], the bottom-up models are considered to be incomprehensive in comparison to the top-down models. This may be due to the methodological foundation of the bottom-up approach that ensures that only a part or unit of the system under consideration is modelled at a time before they are aggregated in order to provide similar information as the top-down approach. While it is unarguably true that the two approaches of top-down and bottom-up represent the two main alternatives to modelling energy consumption and carbon emissions in the domestic sector of economy, Johnston [61] submits that both of them share a degree of commonality. These, according to him, are that (1) they possess the capability to operate at the same level of disaggregation, and (2) they both use the same information, but in different ways. A further discussion of issues regarding the top-down and bottom-up energy models follows.

2.5.2 *Modelling Housing Energy and Carbon Emissions Using the Top-Down Approach*

Based on the foregoing argument, housing energy models using the top-down modelling technique employ an approach that is based on aggregated data and this performs well at national level. Principally, the technique takes the relationships between the energy sector and the large-scale economy in order to develop the model of housing energy and carbon emissions. The approach works well by fitting the historical time series data on housing energy and carbon emissions to macroeconomic variables using econometric and multiple regression methods. Based on the submissions of Kelly [67] and Johnston [61], the technique is capable of giving explanations to the variance between dependent and independent covariates. Among the data that are normally considered when such top-down econometric top-down models are being developed are fuel prices, gross domestic product, income, and average dwelling efficiency.

The top-down approach using the econometric technique has been critiqued within the housing energy studies based on its lack of flexibility, among others. It was argued that the econometric technique finds it difficult to incorporate issues relating to the current and future technological improvement. According to the MIT [79], this is in addition to some other variables influencing housing energy and carbon emissions contrary to using only the macroeconomic time series data. The line of thought of Kelly [67] is in no way different from that of the MIT [79] regarding the criticism of the technique. Kelly [67] contends that there is a lack of details on the best way to incorporate the changes in environmental, social, and economic dimensions to housing energy models based on the climate change challenge by the econometric top-down modelling approach. Among the criticisms received by the top-down modelling approach is its inability to take into consideration the sociotechnical and behavioural aspects of housing energy and carbon emissions at the disaggregated level [86]. As argued in Sect. 2.4.2 of this book based on the submission of Hitchock [44], the issue of housing energy and carbon emissions is a sociotechnical problem that requires solving the problem from both the social science and engineering perspectives.

Within the field of housing energy, the top-down simulation technique has been explored. An example of this can be found in the work of Hirst et al. [43] who used the top-down approach to evaluate the domestic energy consumption sensitivity to demographic, economic, and technological factors. The outcome of the models in the study suggests that the domestic energy use is sensitive to some demographic and economic variables. Those variables are required to be updated regularly in order to ameliorate the quality of the model outputs. Likewise, the work of Haas and Schipper [39: 438] made use of this same technique to explore the role of efficiency improvements on domestic housing energy demand. The output of the research indicates '…a non-elastic response to energy consumption due to irreversible improvements in technical efficiency'.

Further to the above, the work of FitzGerald et al. [31] employs the top-down technique to model energy demand in the Republic of Ireland. The result emanating from the model shows a surge in the value of electricity demand of up to 5% on an annual basis between the years 1960 and 2001. However, the output of the model shows an upward trend in the region of 1.2% annually within the area of non-electricity demand during the same period. It is important to note that the model only considered the effect of cost on energy demand without any recourse to other variables affecting electricity consumption. Additionally, Summerfield et al. [103] use the top-down modelling technique based on multiple regression analysis to model the annual delivered energy price and temperature (ADEPT). The output from the model results in annual household energy consumption in the UK starting from 1970. Accordingly, Lee and Yao [71] contend that the efficacy of the ADEPT model is in its ability to forecast the overall housing energy consumption accurately. However, the ability of the model to forecast the overall housing energy consumption for the short term is limited. Considering these limitations, Swan and Ugursal [104] argue that the top-down modelling technique may be unsuitable in identifying key areas for improvements regarding the demand side of energy consumption at household level.

2.5.3 Modelling Housing Energy and Carbon Emissions Using the Bottom-Up Approach

The bottom-up technique for modelling housing energy and carbon emissions uses disaggregated data with high resolution as input for the simulation as evidenced from the studies of Mhalas et al. [77] and Hoogwijk et al. [46]. According to Shorrock and Dunster [97] and Johnston [61], the application of the bottom-up approach requires input data that are based on quantitative or hard data of physically measurable variables. Examples of these kinds of variables include 'energy efficiency of hot water system' and 'dwellings' fabric insulation in terms of thermal performance', among others. Shorrock and Dunster [97] and Johnston [61] further argue that the disaggregated quantitative data are captured from units such as individual dwellings, groups of dwellings, or households in order to provide the model of housing energy and carbon emissions based on the unit used. An accumulation of the outputs for the units gives the sectorial, regional, or national level of housing energy and carbon emissions as the case may be.

The models developed based on the bottom-up technique vary substantially according to their data requirement and structure. As such, many researchers admit that the bottom-up modelling technique can be categorised into two basic epistemic approaches [61, 64, 67, 71, 77, 104]. The categorisations are given/indicated as the statistics and building physics[1] approaches. In addition to those two epistemic approaches, Kavgic et al. [64] argue the case of combining both the statistical and building physics techniques to produce a hybrid bottom-up modelling method, which is richer and more highly sophisticated. The hybrid bottom-up approach has been demonstrated as evidenced in the Canadian Hybrid Residential End-use Energy and Emission Model (CHREM) [105, 80].

Statistical approach

Many researchers have used the statistical approach of the bottom-up technique extensively within the housing energy and carbon emissions sector. As an example, the approach has been used to simulate and then generate insights regarding housing energy consumption as a function of household characteristics. The statistical approach has created a platform to easily map the energy billing data of occupants of dwellings to household characteristics as collected and made available by energy suppliers. However, these data may not be readily available to the public because of the sensitive information of householders contained therein. Within the body of knowledge, three methods of statistical technique have been identified and documented by different researchers as evidenced in the work of Swan and Ugursal [104]. These techniques are the regression analysis (RA), conditional demand analysis (CDA), and neural network (NN) technique.

According to Fung [32], the model of housing energy and carbon emissions is developed based on the RA approach. The analysis is done in such a way that the housing energy consumption and carbon emissions are regressed on the variables of interest that influence them. The developed models are calibrated based on real-life data and then validated in order to build confidence in the model. As such, the developed models are assessed and evaluated based on some criteria, among which is goodness of fit. Insignificant variables according to the result of the evaluation are removed from the models. However, the use of the CDA approach for housing energy and carbon emissions has been found to be profound. The technique bases

[1]Some researchers use physical or engineering for building physics categorisation.

its method of analysis on the need to regress the household energy consumption on available end-use appliances within the household. The advantage of the technique, according to Swan and Ugursal [104], is premised on the ease of obtaining relevant information required for the model. That is, the information needed for this kind of model can be obtained by conducting a simple survey of occupants' appliances and integrating the collected data with that of energy billing data from the energy suppliers. The main drawback to the approach, however, is that data from many households are required in order to obtain reliable results from the models. The NN approach is based on a simplified neural network mathematical model.

Within the housing energy and carbon emissions research, the statistical methods have gained prominence, particularly when it comes to using disaggregated data at household level. Examples of this research can be traced to the work of Tonn and White [107] who used the RA approach to develop models of electricity consumption for space heating, appliances and lighting, wood use, and indoor temperature. Similarly, the study of Douthitt [26] in Canada uses the RA technique to model the household space heating fuel consumption from historic fuel price, substitute fuel price, total fuel consumption, and a vector of building structure, climatic, and occupants' characteristics. Likewise, Kavousian, Rajagopal, and Fischer [65] apply the RA method to analyse large data sets about the household electricity consumption in order to generate insights regarding the effectiveness of energy efficiency measures.

In addition to the examples cited above relating to the studies that used the RA technique, there are many other studies that attempted the use of the CDA technique in order to produce the statistical bottom-up models regarding housing energy and carbon emissions. Among those studies are the work of Swan and Ugursal [104], Aydinalp et al. [8], Lins et al. [73], Hsiao et al. [47], Goldfarb and Huss [35], Caves et al. [21], Aigner et al. [3], and Parti and Parti [89]. However, the application of the NN approach to model issues relating to housing energy and carbon emissions has been limited. According to Swan and Ugursal [104], the main reason adduced for this is the high computational and data requirements of the technique. That notwithstanding, some studies like those of Yang et al. [116], Aydinalp et al. [9], Mihalakakou et al. [78], Aydinalp et al. [7], and Issa et al. [57] have attempted to use the technique.

Building physics approach

The bottom-up modelling technique based on the building physics approach uses highly disaggregated data for individual dwellings to model housing energy and carbon emissions. According to Swan and Ugursal [104], the approach does not necessarily require any historical data before being used for modelling. Most often, the physical characteristics of dwellings are used in developing models under the building physics approach of the bottom-up technique. As such, mainly quantitative data of physically measurable variables are required in developing housing models [61, 97]. Such information includes details on dwellings' fabric insulation, efficiency of space heating or hot water systems, internal temperatures and heating patterns, external temperatures, and ventilation rates, among others [77]. According to Wilson and Swisher [115], the advantage of using the building physics technique to model the housing energy and carbon emissions over other techniques can be seen in the fact that the building physics approach benefits more from using a combination of data on the physical characteristics of dwellings and other empirically available data from the national database, including the house condition surveys. Over the years, three basic approaches have been used in analysing the housing energy and carbon emissions according to the building physics techniques that have evolved [104]. Those techniques are the distributions, archetypes, and sample techniques.

In the distributions method, the appliances ownership profiles of different households or dwellings within the housing stock are triangulated with the ratings of those appliances. This is mainly to model the likely housing energy consumption and the corresponding carbon emissions based on end-uses in the households or dwellings. Based on this technique, the regional or national energy consumption and carbon emissions can then be estimated by aggregating appliances consumption patterns for each household or dwelling. Contrary to the distribution method, the archetypes approach bases its estimation of energy consumption and carbon emissions on the classification of the housing stock according to dwelling type, size, age, or even tenure. The consumption and emissions patterns for different dwelling type representatives are then summed up before being aggregated to form the regional or national energy consumption and carbon emissions as the case may be. Within the sample technique, however, the approach uses the actual sample dwellings data collected via a survey to develop the model of housing energy and carbon emissions. It is important to note that the actual data collected serve as input to the model. The approach uses a methodology that ensures that the sampling exercise is rigorous and scientifically proven to be the true representative of the population. An example of this approach can be the methodology for sampling for the English or Scottish house condition survey. Using this approach, the consumption and emissions of a variety of dwelling types can be accounted for and then form the basis for the modelling exercise, of which the aggregated result constitutes the regional or national estimate.

The distributions, archetypes, and sample methods of building physics approach of the bottom-up modelling technique have been applied in many studies to model housing energy and carbon emissions. For instance, both the developing and developed countries have made use of the distributions technique to model housing energy and carbon emissions at both regional and national levels. An example of this can be demonstrated by the work of Saidur et al. [93] who applied the distributions method of appliances ownership to model a non-space heating household energy in Malaysia. The result of the model gives the annual energy consumption for the entire country. Further to the above, the study of Kadian et al. [62] developed a model of energy-related emissions for households in Delhi, India using a combination of the distributions and micro-level data sources. Similarly, the study of Capaso et al. [20] in Italy made use of the appliance usage profile of householders according to the distributions technique in order to give an outlook of energy consumption for the entire housing stock in that country. The input to the model is a combination of the householders' lifestyle data and engineering data of different types of appliances. Likewise, the study of Jaccard and Baille [58] in Canada applied the distributions technique to model carbon emissions' reduction cost of householders based on appliances use behaviour of the householders.

The archetypes technique of the building physics modelling method has been extensively utilised by many modellers within the housing energy and carbon emissions. Consequently, many publications have emerged within this domain in the body of literature. For example, the model of Parekh [88] in Canada was developed based on archetypes of dwellings' characteristics in a bid to simplify the analysis and evaluation of household energy use. Huang and Broderick's [48] studies were conducted in the USA. The study modelled space heating and cooling loads of the USA housing stock for 16 different regions using 16 multi-family and 45 single-family archetypes of dwellings. The results of the study produced an energy consumption profile for space heating and cooling loads of 16 different dwelling archetypes for the entire USA housing stock. One of the most profound insights from the output of the model is the way the results were disaggregated in such a way that the contributions of thermal conductivity of walls, roof, windows, and others could be evaluated.

Further to the studies above, the work of Petersdorff et al. [90] uses three different archetypes of dwellings (terrace, small apartment, and large apartment) for modelling the European Union (EU)-15 building stock. The study examines and considers five standard dwellings and eight insulation standards using the built environment analysis model. The results produced the heating demand based on the archetypes for 15 different EU countries. Similarly, the work of Johnston [61] focused on the UK and developed the energy consumption and carbon emissions model for the UK housing stock to represent different types of dwellings. This model was also disaggregated into two types of dwellings based on construction date (i.e. pre-1996 and post-1996). The disaggregation according to this model reflects the entire UK housing stock. There are a number of other studies that have used the archetypes approach for their models. Among those studies are those of Clarke et al. [23], Jenkins [59], Gupta [38], Firth et al. [30], Natarajan et al. [85], and Mhalas et al. [77].

Comparing the sample method to the archetypes approach, the sample method as one of the techniques of building physics modelling approach has been sparingly applied. This may be due to the huge amount of data required by the approach. As such, the approach has not been reported much in the literature. Among the few studies that reported the approach are those of Shorrock and Dunster [97], Farahbakhsh et al. [29], Larsen and Nesbakken [70], Boardman et al. [17], and Natarajan and Levermore [83]. The Building Research Establishment's Housing Model for Energy Studies (BREHOMES) [97] developed in the early 1990s used 1000 dwelling types (defined by age group, built form, tenure type, and ownership of central heating) as the sample upon which the annual household energy consumption of UK housing stock is based. A total of 8787 dwellings (defined by type, space heating fuels, vintage, and province) were used in Canada [29] to provide the

Canadian Residential Energy End-use Model (CREEM) in a bid to test the effect of different strategies of carbon reductions based on two standards.

The model developed by Larsen and Nesbakken [70] used 2013 dwellings to produce the model of household energy consumption of Norway's housing stock. The UK domestic carbon model (UKDCM) developed by the Environmental Change Institute of Oxford University [17] made use of 20,000 dwelling types using national statistics to produce the monthly housing energy and carbon emissions profiles. The model produced three different scenarios until 2050. In concluding this section, the domestic energy and carbon (DECarb) model [83] developed by the research unit for the Engineering and Design of Environments in the Department of Architecture and Civil Engineering at the University of Bath in 2007 used 8064 unique combinations for six age bands of the UK housing stock to produce the monthly energy consumption for the UK.

2.5.4 Benefits and Limitations of the Top-Down and Bottom-Up Modelling Approaches

Stemming from the above, it is important to note that both the top-down and bottom-up approaches have their strengths and drawbacks. It is therefore imperative to have them summarised before critically reviewing some extant energy models within the housing sector of the UK economy. As such, Table 2.2 summarises some benefits and limitations of the top-down and bottom-up approaches as evidenced from the work of Kavgic et al. [64]. It is necessary to add that both the top-down and bottom-up approaches have not utilised the advantage of incorporating qualitative data input. However, Kavgic et al. [64] fails to recognise this as one of the limitations of those approaches. This then limits the capability of those approaches. Importantly, incorporating both the qualitative and quantitative data sources to any modelling efforts improves the robustness of such approaches as its lack undermines their capabilities.

2.6 Extant Energy and Carbon Emissions Models Within the UK Housing Sector

From Sect. 2.5 of this research book, it is obvious that several models have evolved in the UK over the years to estimate and forecast the current and future profiles of energy and carbon emissions within the UK housing stock. It is therefore important to discuss in detail some of the housing energy and carbon emissions models that are specifically developed for the UK. Among those models are the following:

Table 2.2 Benefits and limitations of top-down and bottom-up modelling approaches

Characteristics	Top-down	Bottom-up statistical	Bottom-up building physics
Benefits	– Focus on the interaction between the energy sector and the economy at large – Capable of modelling the relationships between different economic variables and energy demand – Avoid detailed technological descriptions – Able to model the impact of different social cost-benefit energy and emission polices and scenarios – Use of aggregated economic data	– Include macroeconomic and socioeconomic effects – Able to determine a typical end-use energy consumption – Easier to develop and use – Do not require detailed data (only billing data and simple survey)	– Describe current and prospective technologies in detail – Use physically measurable data – Enable policy to be more effectively targeted at consumption – Assess and quantify the impact of different combination of technologies on delivered energy – Estimate the least-cost combination of technological measures to meet given demand
Limitations	– Depend on past energy economy interactions to project future trends – Lack the level of technological detail – Less suitable for examining technology-specific policies – Typically assume efficient gaps	– Do not provide much data and flexibility – Have limited capacity to assess the impact of energy conservation measures – Rely on historical consumption data – Require large sample – Multi-collinearity	– Poorly describe market interactions – Neglect the relationships between energy use and macroeconomic activity – Require a large amount of technical data – Do not determine human behaviour within the model but by external assumptions

Adapted from Kavgic et al. [64]

- The BREHOMES [97, 98].
- The Johnston model [61].
- The UKDCM [17].
- The DECarb model [83].
- The Community Domestic Energy Model (CDEM) [30].
- The Cambridge Housing Model (CHM) [49, 48].
- The Domestic Dwelling Model (DDM) [77].

2.6.1 The Building Research Establishment's Housing Model for Energy Studies

The Building Research Establishment's Housing Model for Energy Studies (BREHOMES) as developed by Shorrock and Dunster [97] is seen as the earliest UK housing energy model that is based on the building physics method of the bottom-up modelling approach to estimate the energy and carbon emissions for the UK housing stock. The model is highly disaggregated and uses the weighted average stock transformation method to convert over 18,000 households surveyed to over 1000 different dwelling types in a bid to build this dwelling types profile. The core calculation engine for the model is based on the Building Research Establishment's Domestic Energy Model (BREDEM) in order to establish energy use for dwellings. The BREDEM energy model accepts input on different areas of dwelling elements to include dwellings' thermal characteristics, number of occupants or household size, internal and external temperatures, solar gains, and heating patterns, among others.

The output of the model produces the annual energy consumption and carbon emissions at the national level of aggregation. Two different scenarios are explored by the model to include (1) the baseline model termed the 'Reference' (business-as-usual) scenario and (2) 'Efficiency' scenario. The earlier version of this model as reported in Shorrock and Dunster [97] produces the output from a base year of 1990 to 2020. However, a more recent version of the model as reported in Shorrock et al. [98] used a base year of 1993 and extends the output trends till 2050. The model has been extensively applied as a policy advice tool for the Department for Environment, Food, and Rural Affairs (DEFRA). However, the model is not transparent as it is difficult to replicate the study. Also, the model lacks the capability of capturing qualitative data as input data source as it relies heavily on highly disaggregated quantitative data source.

2.6.2 The Johnston Model

The Johnston model is another one of the notable energy and carbon emissions models for the UK housing stock. As in the case of the BREHOMES, it is also a model based on the building physics technique of a bottom-up modelling approach. The Johnston model has the capability of reflecting the different types of dwellings of the entire UK housing stock. However, the model basically disaggregated the overall housing stock into two by using dwellings' year of construction as the main criterion for the disaggregation. Here, the entire UK housing stock is represented in the model as (1) pre-1996 and (2) post-1996. Similar to the BREHOMES, the model adapted the BREDEM's calculation algorithm for each dwelling type in order to calculate energy and emissions of these individual dwelling types.

The overall output of the model accurately produces the total annual energy consumption and carbon emissions for the entire UK housing stock with 1996 as the base year and this continues until 2050. This is in a bid to give the previous and current household consumption and emissions level(s), and then project into the future regarding the household consumption and emissions level. In order to explore the effects of changes to certain assumptions (such as the uptake of new technology, trends in population, and energy usage changes, among others.) regarding the model, three major scenarios as typical applicability of the model were produced. The first scenario, termed the 'business-as-usual (BAU)', looks at the current trends and projects these until 2050 with an assumption that there will not be any further action or intervention from government to reduce the emissions. With these trends, the output of the scenario reveals that about 33% of the emissions could be reduced by the year 2050 when compared to the emissions' level of 1996, which was used as the base year.

The second scenario, termed the 'demand side', is based on BAU and extends it in order to incorporate some other measures should new evidence regarding the climate change, for example, emerge in the near future. This scenario explores improvements in the energy efficiency of the demand side of household energy. The output of the model predicted a 58% reduction in carbon emissions for this scenario. Additionally, the third scenario, termed the 'integrated' scenario, combines both the supply and demand sides of the UK housing stock and explores their effects on the carbon emissions of the entire housing stock. The results show that an approximate 74% reduction is achievable. Johnston [61] notes the limitations of the model, namely that the model suffers from usability and transparency and he recommends transferring the model to a more suitable platform. Regarding the application of the model, it is capable of being used as a policy advice tool. However, the application of the model has been limited to its developer alone. Therefore, it has not been extensively used in practice.

2.6.3 The UK Domestic Carbon Model

The UKDCM model was developed by the Environmental Change Institute of the Oxford University in 2005 in order to explore and investigate how a 60% reduction in carbon emissions could be achieved in the UK housing sector. The model is based on building physics. The model processes a huge amount of data that include those obtained from the English Housing Condition Survey and its equivalents in Scotland and Northern Ireland. Among the data input required by the model are population figures, levels of insulation, and the efficiency of heating equipment, among others, as contained in the 40% House report [17]. The model contains highly disaggregated data sets for nine geographical areas, seven age classes, and ten types of construction of some 20,000 UK dwellings. Additionally, the model has the capability to process different combinations of these data sets in order to further subdivide the dwellings based on tenure, construction method, and number of floors. Similar to the BREHOMES and Johnston models, the model made use of the BREDEM calculation engine to estimate the emissions of these dwellings.

The output and temporal resolution of the model give the monthly energy consumption and carbon emissions of the UK housing stock. The model performed three different scenarios to explore the effects of some policy formulations regarding energy use. The scenarios tested are (a) BAU, (b) 44% emission reduction, and (c) 25% emission reduction below 1990 levels. This model, together with the scenarios tested, was improved upon by another version of the model termed UKDCM2 [42]. According to Boardman [18], this newer version was used to prepare the Home Truths report, where the analysis of different scenarios regarding a reduction in future carbon emissions was conducted and explored. The scenario A of this newer version represented a plausible scenario to reflect what would happen should there be '*…a continuation of current and near-terms trends, technologies, policies and practices, with changes occurring slowly into the future*' [42]. Scenario B updates the scenario B of the '40% House report' and this now investigates the way government's target of 60% emissions' reduction by the year 2050 could be achieved through the assumption that members of society now know more about the issue of energy use and carbon emissions with attendant technological and societal changes to bring about a reduction in carbon emissions. On the other hand, scenario C explores how a further reduction in carbon emissions in excess of 60% could be achieved by assuming a higher uptake of renewable and other efficient energy sources, additional demolition and new build, among others. The model is being used generally as a policy advice tool and it is freely available over the Internet.

2.6.4 The Domestic Energy and Carbon Model

The DECarb is another notable model of the energy and carbon emissions models for the UK housing stock with the capability of mapping different technical and climate scenarios in order to generate future trends and options regarding consumption and emissions. The model is an object-oriented one that is capable of running on any of the operating systems and it is user friendly in terms of selecting input data. The model is based on the building physics approach according to the work of Natarajan and Levermore [83]. The DECarb model uses a highly disaggregated housing stock approach that has 8064 unique combinations of six historical age classes of the UK housing stock. Like other previously discussed models, DECarb made use of the BREDEM algorithm for the calculation of consumption and emissions profiles for individual dwellings in the model. There are six different files for the data set with each representing each of the six different age classes and this data set consists of different variables that include dwelling type, and insulation characteristics, among others. The model is then used to run future scenarios regarding UK housing stock.

As the overall output, the model calculates the annual energy consumption and carbon emissions in a bid to perform a forecast of their trends from the base year 1996 until 2050. Interestingly, the model has the capability of performing a back-cast analysis from 1996 backwards. This is embedded into the model in order to serve as a way of validating the model. The model was then used to test climate change scenarios according to UKCIP02 in addition to the BREHOMES, Johnston, and UKDCM scenarios. For example, using the Johnston's model scenarios, the results suggest that it is unlikely to meet the target of a reduction of up to 50% in carbon emissions for all these scenarios run [84]. As for other models discussed above, the model is being used as a policy advice tool and is readily available online as open framework. As noted above, the model is user friendly in selecting the input data; however, the mode of output data presentation is poor as they are displayed in text file. This then presents difficulty in reading the results of the model.

2.6.5 The Community Domestic Energy Model

The Community Domestic Energy Model (CDEM) is another notable model of energy consumption and carbon emissions of the UK housing stock that was developed by the Department of Civil and Building Engineering, at Loughborough University in 2009 [30] based on the building physics approach. In like manner to other previously mentioned models, this model is highly disaggregated, but with 47 house archetypes that are derived from unique combinations of built form type and dwelling age. For the house architecture calculation engine, the model requires input from many sources including the English House Condition Survey (EHCS), the BREDEM-8 calculation engine, and the SAP rating, among others.

Regarding the core dwelling model, the main data requirement comes from the BREDEM-8 calculation engine, monthly average external temperatures and monthly average solar radiation as obtainable from the Met Office. The output of the model produces monthly energy consumption and carbon emissions of the whole UK housing stock. Also, the model is capable of producing output based on city or neighbourhood housing stock. Apparently, the model fails to test any scenarios; instead the model is used to estimate and predict energy consumption and carbon emissions of the 2001 English housing stock alone.

2.6.6 The Cambridge Housing Model

The Cambridge Housing Model (CHM) was developed by the Cambridge Architectural Research in a bid to forecast energy consumption and carbon emissions for housing stock in the UK in general. It is another building physics-based bottom-up model that uses the calculations formulated and established by the SAP

2009 [19] and BREDEM engine [97] in order to perform all its internal calculations. The model has three basic data input components which include climate data, housing data, and building physics data. For climate data input, the model uses SAP's monthly solar declination and regional latitude data, BREDEM-8's monthly/regional solar radiation data, monthly or regional year-specific wind speed and external temperature data as taken from a number of different stations across the UK.

Regarding the housing data input, the main source here is based on 16,670 dwellings as contained in the English Housing Survey of 2010 [87] with an adjustment to scale this up to reflect the UK housing stock. However, the building physics data input is the direct result of the calculations performed in the SAP and BREDEM. The model then reads in data for individual representative dwellings in order to perform building physics calculations. The CHM is one of the most transparent models because the model is built and all its calculations performed in Microsoft Excel.

The output of the model therefore gives the energy consumption together with associated carbon emissions according to fuel and end-use. These are presented as representative of each dwelling type, English housing stock, and Scottish housing stock, among others, as well as for the entire UK housing stock. It is worthy of note that the output of this model is one of the studies that made up the UK housing fact file domicile in the Department of Energy and Climate Change (DECC) [87].

2.6.7 The Domestic Dwelling Model

The Domestic Dwelling Model (DDM) is a new approach being proposed by the Technology Futures Institute of the Teesside University [77] to model energy consumption and carbon emissions of dwellings and neighbourhoods based on visualisation. The model is highly disaggregated as it estimates each dwelling independently within the neighbourhood. The model uses the SAP/BREDEM energy calculation engine. As input to the SAP core calculation engine, the model utilises information from aerial and terrestrial imagery, digital maps, household surveys, census, and the Office for National Statistics (ONS). As a first step in the development of this approach, models of the dwellings in the neighbourhood are developed [77].

Following this is the undertaking of the energy performance calculation of the models according to the SAP algorithms. The carbon emissions' reduction capability of the dwelling is therefore quantified based on the existing characteristics of the dwelling before using a decision support system to implement the effectiveness of energy improvement measures. This model is implemented on a geographic information system (GIS) platform.

2.7 Critique of the Modelling Methods

By considering the contribution of emissions from the domestic dwellings, it can be deduced from Sect. 2.6 that considerable efforts have been invested into energy models for dwellings in the UK. One thing that is common and central to all the reviewed models under Sect. 2.6 of this research book is that they all share the same BREDEM algorithms in estimating and forecasting energy consumption and carbon emissions. BREDEM has been adjudged as a well-established method to accurately predict UK dwelling energy consumption [85] as it forecasts dwellings' energy consumption and carbon emissions at a highly disaggregated level based on deterministic building physics. Additionally, the domain of application of these models is common as they are all applied as policy advice tools. However, the models are varied in terms of their level of disaggregation, resolution of output, output aggregation level, scenario analysis performed, model validation, and their availability to the members of public for scrutiny, as shown in Table 2.3.

Regrettably, these models have been criticised owing to a number of limitations. Firstly, all the models have been criticised for their low level of transparency. Kavgic et al. [64] and Mhalas et al. [77] argue that the models' transparency, in terms of the architecture and data sources, is seen as one of the most essential issues worth considering for future deployment of the models. However, some of these models are not available to the members of public; even those that are made available to public contain little information about their structure and operational details. As such, the models could not be scrutinised as obtaining access to raw input data or the algorithms used by the models has been possible for the majority of them. This is because it is unclear on how the relationships among the different variables making up the models are formulated and built up. Consequently, the outputs of these models are extremely difficult to replicate.

Secondly, the models fail to take into consideration the complex inter-dependencies and dynamic nature of the issue of energy consumption and carbon emissions, especially in households. This is because the modelling approaches of these models are based on a static and deterministic method, which is classified as a reductionist paradigm that uses linear orientation to give the forecast of a system, which, for example, is just for a particular point in time. These models therefore work with particular sets of data inputs in a bid to produce particular sets of outputs that have little or no room to accommodate uncertainty in input data sets. This is because the approaches for the models are hinged on the notion that exact relationships exist without uncertainty between the variables in the models. For example, some of them employ the use of simple regression analysis that relies on historical data. Here, the future trends are predicted based on the historical without taking into consideration any undesirable or chaotic events that may occur in the near future.

Thirdly, the importance of occupants' dwelling interaction cannot be over-emphasised regarding energy consumption in homes. Therefore, special attention needs to be accorded to this aspect as well. This is evidenced from the

Table 2.3 Comparative analysis of notable UK energy models

Characteristics models	Developer	Year	Calculation model embedded	Output resolution	Disaggregation level	Scenario analysis	Data output aggregation level	Validation with historical data	Application	Availability
BREHOMES	Building Research Establishment (BRE)	Early 1990s	BREDEM-12	Annual energy consumption	1000 dwelling types used based on age group, tenure type, etc.	Two scenarios until 2020: (a) Reference (b) Efficiency	National	Extensive	Policy advice tool	Available to the developer only
Johnston	Johnston's Ph. D. thesis at Leeds University	2003	Modified version of BREDEM-9	Annual energy consumption and carbon emissions	Two dwelling types separated into pre- and post-1996	Three scenarios until 2050: (a) Business-as-usual (b) Demand (c) Integrated	National	Comparison with historical data from the BRE domestic energy fact file	Policy advice tool	Available to the developer only
UKDCM	Environmental Change Institute (ECI), Oxford University	2006	BREDEM-8	Monthly energy consumption and carbon emissions	20,000 dwelling types by 2050	Three scenarios until 2050: (a) Business-as-usual (b) 44% emission reduction (c) 25% emission reduction below 1990 levels	National	Validated with regional historical data provided by BERR	Policy advice tool	Open
DECarb	Research Unit for the Engineering and Design of Environments, Department of Architecture and Civil Engineering, University of Bath	2007	BREDEM-8	Monthly energy consumption	8064 unique combinations of six age bands of different dwelling types	(a) UKIPO2 scenarios (b) BREHOMES scenarios (c) Johnston scenarios (d) UKDCM scenarios (e) Back-cast scenario (1970–1996)	National	Validated with historical data provided by BERR and BRE domestic energy fact file	Policy advice tool	Open

(continued)

Table 2.3 (continued)

Characteristics models	Developer	Year	Calculation model embedded	Output resolution	Disaggregation level	Scenario analysis	Data output aggregation level	Validation with historical data	Application	Availability
CDEM	Department of Civil and Building Engineering, Loughborough University	2009	BREDEM-8	Monthly energy consumption and carbon emissions	47 dwelling archetypes based on unique combinations of age band and dwelling built form types	A scenario to predict 2001 housing stock	National, city, neighbourhood	Validated with 2001 DEFRA aggregate figure of domestic space heating consumption	Policy advice tool	Open
CHM	Cambridge Architectural Research, University of Cambridge	2010	BREDEM-8 and SAP 2009	Annual energy consumption and carbon emissions	16,670 dwelling types	Conducted sensitivity analysis for the 15 most sensitive parameters in the model	National, regional	Extensive	Policy advice tool	Open
DDM	Technology Futures Institute, Teesside University	2013	BREDEM-8 and SAP 2009	Annual energy consumption and carbon emissions	756 dwelling types	Illustrative scenarios of (a) Fabric change (b) Solar PV (c) μ-CHP (d) Condensing boiler (e) ASHP (Under-floor) (f) ASHP (radiator)	Individual, neighbourhood	Validated with energy performance certificate of some social housing dwellings in Middlesbrough	Policy advice tool	Available to the developer only

Adapted from Oladokun and Odesola [86]

assertion made on the report of the Inter-Governmental Panel on Climate Change (IPCC) mitigation [53] that '...*occupant behaviour, culture and consumer choice and use of technologies are also major determinants of energy use in buildings and play a fundamental role in determining carbon emissions*'. Conversely, the IPCC [53] report further suggests that there is limited evidence to show that these determinants are being incorporated into energy models. While it is evident that BREDEM, for example, incorporates some degree of occupants' aspect such as the number of occupants into their model, Natarajan et al. [85] confirm that the behavioural aspect has been limited and not explicitly considered.

Fourthly, it is evident that the issue of energy and carbon emissions remains increasingly complex and difficult to manage. This is due to the fact that a number of issues regarding the energy sector of the economy are evolving on a daily basis. For example, in order to accurately predict and forecast energy consumption and carbon emissions, the energy sector would undoubtedly interact with other sectors such as the economic and environment sectors and a host of others. These sectors are difficult to manage on their own merit. However, dynamically integrating these external sectors to the energy sector further compounds the problem of household energy issues. As such, none of the models reviewed in Sect. 2.2.6 have demonstrated enough capacity to dynamically accommodate additional systems that utilise both the quantitative and qualitative data inputs and where some variables may interact in a nonlinear way. This then portends to mean that the models are profoundly limited for their lack of ability to incorporate the feedback from these external sectors.

From the foregoing, it is apparent that there is a need to look both inwardly and outwardly for sophisticated modelling approaches capable of dealing with the limitations above and then model the kind of complexity and challenges that are facing the housing energy and carbon emissions modelling. This may mean further broadening the scope and level of interaction of different housing energy and carbon emissions drivers and the capacity to expand this should the need arise in the near future. In order to further reinforce this, several researchers such as Hitchcock [44], Kohler and Hassler [69], Shipworth [95, 96], and Motawa and Banfill [82] advocate the use of the STS as an approach to model this complexity owing to high inter-dependencies, chaotic, and nonlinearity of the variables involved. It is necessary to emphasise that STS is one of the methodologies of the system-based approach of scientific inquiry. This methodology has previously been used as an approach to model the complexity of real systems' elements and relationships (as will be discussed in Chap. 3). Modelling complexity enables capturing the interdependent and multi-causal correlation structure of the elements of STS and determining the efficacy of different change strategies. This helps in analysing the nonlinear behaviour of the studied systems where changes in input are neither proportional to changes in output, nor is the input-to-output relationship fixed over time. This research book will then use the STS approach to model the housing energy and carbon emissions. The theoretical backgrounds and the modelling techniques for the STS are covered in Chap. 3 of this book.

2.8 Summary

This chapter has shown that there is a wide range of frameworks that previous studies have formulated to conceptualise the issue of energy consumption and carbon emissions, which now serve as the theoretical backgrounds underpinning energy models. They therefore principally fall within two major domains: disciplinary and integrated frameworks. The disciplinary framework focuses on how individual disciplines illustrate the approach to solving energy and carbon emission problems by formulating a framework. For example, the engineering approach looks at the technology of energy consumption and carbon emissions. On the hand, the integrated framework uses a holistic approach to combine a number of disciplines and provide a framework capable of shaping the issue of energy consumption and carbon emissions based on the limitations of the disciplinary framework. Within the body of literature, it was established that the sociotechnical variables influencing housing energy and carbon emissions basically come from three interacting systems comprising the dwellings, the occupants, and the external environment. The variables identified within the dwellings system are related to the dwellings' physical characteristics. Also, variables related to the occupants' system are in terms of household characteristics, occupants' thermal comfort, and occupants' behaviour. Finally, the variables related to the external environment system are in terms of climatic, economic, and cultural influences. The variables used for model conceptualisation in Chap. 5 of this book are drawn from these frames of variables and mapped into six different modules.

Furthermore, the chapter has demonstrated that over the years a number of energy and carbon emissions models have evolved with the capability of forecasting and estimating housing energy consumption and carbon emissions. These models are found to vary considerably based on the levels of disaggregation, complexity, resolution of output, output aggregation levels, scenario analysis performed, model validation, and their availability to the members of public for scrutiny. Basically, two major epistemic approaches have been used that include: top-down or bottom-up approaches. The top-down techniques rely on the kind of interaction between the energy sector and the economy in general at aggregated level in order to predict the behaviour of energy consumption and carbon emissions, especially at the household level, when some changes are made to the policy parameters within such models. On the other hand, bottom-up techniques mainly focus on the energy sector alone, utilising a disaggregated approach of either a statistical or building physics method that contains a high level of detail to model energy consumption and carbon emissions, especially at household level.

After a careful appraisal of the existing modelling approaches, the chapter concludes that there are a number of limitations in the existing modelling techniques that prevent them from being used in this research book. These are (1) lack of transparency in the model algorithms, (2) inability to account for the complex inter-dependencies and dynamic nature of the issue of energy consumption and carbon emissions, (3) limited evidence to show for the occupants' dwelling interactions, and (4) lack of enough capacity to accommodate qualitative data input. As such, there is the need to search for more robust and sophisticated modelling approaches that take into consideration the kind of complexity involved that bedevil the issue of housing energy and carbon emissions owing to high inter-dependencies, chaotic nonlinearity, and the qualitative nature of some of the variables involved.

References

1. Abrahamse W, Steg L (2011) Factors related to household energy use and intention to reduce it: the role of psychological and socio-demographic variables. Hum Ecol Rev 18 (1):30–40
2. Accenture (2010) Understanding consumer preferences in energy efficiency. Available at http://www.accenture.com/SiteCollectionDocuments/PDF/Understanding_Consumer_Preferences_Energy_Efficiency_10-0229_Mar_11.pdf. Accessed 15 July 2011
3. Aigner D, Sorooshian C, Kerwin P (1984) Conditional demand analysis for estimating residential end-use load profiles. Energy J 5(4):81–97
4. Ajzen I (1985) From intentions to actions: a theory of planned behaviour. In: Kuhl J, Beckman J (eds) Action control: from cognition to behaviour. Springer, Berlin, pp 11–39
5. Ajzen I (1991) The theory of planned behaviour. Organ Behav Hum Decisi Process 50 (2):179–211
6. Anderson BR, Clark AJ (1985) BREDEM: BRE domestic energy model: background, philosophy and description. Building Research Establishment, Watford
7. Aydinalp M, Ugursal VI, Fung AS (2002) Modelling of the appliance, lighting, and space cooling energy consumptions in the residential sector using neural networks. Appl Energy 72(2):87–110
8. Aydinalp M, Ugursal VI, Fung AS (2003) Modelling of residential energy consumption at the national level. Int J Energy Res 27(4):441–453
9. Aydinalp M, Ugursal VI, Fung AS (2004) Modelling of the space and domestic hot-water heating energy-consumption in the residential sector using neural networks. Appl Energy 79 (2):159–178
10. Azar E, Menassa CC (2012) Agent-based modelling of occupants and their impact on energy use in commercial buildings. J Comput Civ Eng 26(4):506–518
11. Baba A, Mahjoubi L, Olomolaiye P, Colin B (2012) Insights of architects' knowledge of the code for sustainable homes in relation to low carbon housing design and delivery in the UK. Struct Surv 30(5):443–449
12. Baker P (1991) A model of household gas and electricity expenditures for the UK: the IFS simulation program for energy demand (SPEND). Institute for Fiscal Studies, London
13. Barr S, Gilg A (2006) Sustainable lifestyles: framing environmental action in and around the home. Geoforum 37:906–920
14. Bartiaux F, Gram-Hanssen K (2005) Socio-political factors influencing household electricity consumption: a comparison between Denmark and Belgium. In: ECEEE 2005 summer study —what works and who delivers, pp 1313–1325
15. Bin S, Dowlatabadi H (2005) Consumer lifestyle approach to US energy use and the related CO_2 emissions. Energy Policy 33:197–208
16. Böhringer C, Rutherford TF (2008) Combining bottom-up and top-down. Energy Econ 30:574–596
17. Boardman B, Darby S, Killip G, Hinnels M, Jardine CN, Palmer J (2005) 40% house. University of Oxford's Environmental Change Institute, Oxford, UK
18. Boardman B (2007) Home truths: a low carbon strategy to reduce UK housing emissions by 80% by 2050. Environmental Change Institute, University of Oxford, UK

19. Building Research Establishment (BRE) (2011) The government's standard assessment procedure for energy rating of dwellings. 2009 edition incorporating RdSAP 2009, Watford, UK
20. Capaso A, Grattieri W, Lamedica R, Prudenzi A (1994) A bottom-up approach to residential load modelling. IEEE Trans Power Syst 9(2):957–964
21. Caves DW, Herriges JA, Train KE, Windle RJ (1987) A Bayesian approach to vombining conditional demand and engineering models of electricity usage. Rev Econ Stat 69(3):438
22. Chartered Institution of Building Services Engineers (CIBSE) (2013) The limits of thermal comfort: avoiding overheating in European buildings (CIBSE TM52). CIBSE, London
23. Clarke JA, Ghauri S, Johnstone CM, Kim JM, Tuohy PG (2008) The EDEM methodology for housing upgrade analysis, carbon and energy labelling and national policy development. In: Proceedings of the IBPSA Canada, eSim conference, Quebec City, Canada, pp 135–142
24. Department of Energy and Climate Change (2012) Low carbon community challenge report. Available at https://www.gov.uk/government/uploads/system/uploads/attachment_data/file/48458/5788-low-carbon-communities-challenge-evaluation-report.pdf. Accessed 9 May 2012
25. Department of Trade and Industry - DTI (2003). Energy white paper 2003 – our energy future: creating a lowcarbon, TSO (The Stationery Office), Norwich
26. Douthitt RA (1989) An economic analysis of the demand for residential space heating fuel in Canada. Energy 14(4):187–197
27. Elforgani MS, Rahmat I (2010) An investigation of factors influencing design team attributes in green buildings. Am J Appl Sci 7(7):976–986
28. Energy Systems Research Unit (2012) Household energy upgrade manual. Available at http://www.esru.strath.ac.uk/Programs/EEff/index.htm. Accessed 9 May 2012
29. Farahbakhsh H, Ugursal VI, Fung AS (1998) A residential end-use energy consumption model for Canada. Int J Energy Res 22(13):1133–1143
30. Firth SK, Lomas KJ, Wright AJ (2010) Targeting household energy efficiency measures using sensitivity analysis. Build Res Inf 38(1):24–41
31. FitzGerald J, Hore J, Kearney I (2002) A model for forecasting energy demand and greenhouse gas emissions in Ireland. Working paper no 146, The Economic and Social Research Institute, Dublin
32. Fung AS (2003) Modelling of national and regional energy consumption and associated greenhouse gas emissions. Ph.D. thesis, Dalhousie University, Halifax
33. Gatersleben B, Steg L, Vlek C (2002) Measurement and determinants of environmentally significant consumer behavior. Environ Behav 34:335–362
34. Gill ZM, Tierney MJ, Pegg IM, Allan N (2010) Low-energy dwellings: the contribution of behaviours to actual performance. Build Res Inf 38(5):491–508
35. Goldfarb DL, Huss R (1988) Building utility scenarios for an electric utility. Long Range Plan 21(2):78–85
36. Gram-Hanssen, K. (2014). New needs for better understanding of household's energy consumption – behaviour,lifestyle and practices? Architectural Engineering and Design Management, 10(1-2), 91-107.
37. Greening LA (1995) Use of region, life-cycle and role variables in the short-run estimation of the demand for gasoline and miles travelled. Appl Econ 27(7):643–656
38. Gupta R (2009) Moving towards low-carbon buildings and cities: experiences from Oxford, UK. Int J Low-Carbon Technol 4:159–168
39. Haas R, Schipper L (1998) Residential energy demand in OECD-countries and the role of irreversible efficiency improvements. Energy Econ 20(4):421–442
40. Harris DJ (2012) A guide to energy management in buildings. Spon Press, London
41. Hart M, De Dear R (2004) Weather sensitivity in household appliance energy end-use. Energy Build 36:161–174
42. Hinnells M, Boardman B, Layberry R, Darby S, Killip G (2007) The UK housing stock 2005 to 2050: assumptions used in scenarios and sensitivity analysis in UKDCM2. Environmental Change Institute, University of Oxford, UK

43. Hirst E, Lin W, Cope J (1977) A residential energy use model sensitive to demographic, economic, and technological factors. Quart Rev Econ Financ 17(2):7–22
44. Hitchcock G (1993) An integrated framework for energy use and behaviour in the domestic sector. Energy Build 20:151–157
45. Hoes P, Hensen JLM, Loomans MGLC, De Vries B, Bourgeois D (2009) User behavior in whole building simulation. Energy Build 41:295–302
46. Hoogwijk M, Van Vuuren DP, Scrieciu S (2008) Sectorial emission mitigation potentials: comparing bottom up and top down approaches. Available at http://www.env.go.jp/press/file_view.php?serial=12478&hou_id=10316. Accessed May 2012
47. Hsiao C, Mountain C, Illman KH (1995) A Bayesian integration of end-use metering and conditional-demand analysis. J Bus Econ Stat 13(3):315
48. Huang Y, Broderick J (2000) A bottom-up engineering estimate of the aggregate heating and cooling loads of the entire US building stock. Lawrence Berkeley National Laboratory, Report LBNL-46303
49. Hughes M (2011) A guide to the Cambridge housing model. DECC, London
50. Hughes M, Palmer J (2012) The Cambridge housing model. DECC, London
51. International Energy Agency (IEA) (1998) Mapping the energy future: energy modelling and climate change policy. Paris
52. International Energy Agency (IEA) (2012) World energy outlook 2012. IEA, Paris, France
53. Intergovernmental Panel on Climate Change (IPCC) (2007) Climate change 2007: mitigation. In: Metz B, Davidson O, Bosch P, Dave R, Meyer L (eds) Contribution of working group III to the fourth assessment report of the intergovernmental panel on climate change, Technical report, IPCC
54. Ironmonger DS, Aitken CK, Erbas B (1995) Economies of scale in energy use in adult-only households. In: Stevens P (ed) The economics of energy, vol 1, pp 318–327. Elgar reference collection. International library of critical writings in economics, vol 119
55. Ironmonger DS, Aitken CK, Erbas B (1995) Economies of scale in energy use in adult-only households. Energy Econ 17(4):301–310
56. Isaacs N, Saville-Smith K, Camilleri M, Burrough L (2010) Energy in New Zealand houses: comfort, physics and consumption. Build Res Inf 38(5):470–480
57. Issa RRA, Flood I, Asmus M (2001) Development of a neural network to predict residential energy consumption. In: Proceedings of the sixth international conference on the application of artificial intelligence to civil & structural engineering computing, held in Eisenstadt, Austria, from 19 to 21 Sept 2001, pp 65–66
58. Jaccard M, Baille A (1996) CO_2 emission reduction costs in the residential sector: behavioural parameters in a bottom-up simulation model. Energy J 17(4):107–135
59. Jenkins D (2008) Energy modelling in traditional Scottish houses (EMITSH): Heriot-Watt University analysis of potential CO_2 savings of building variants. Tarbase Group Report for Historical Scotland—Technical paper, Nov 2008
60. Jennings M, Hirst N, Gambhir A (2011) Reduction of carbon dioxide emissions in the global building sector to 2050. Report GR3. Grantham Institute for Climate Change, Imperial College London
61. Johnston D (2003) A physically based energy and carbon dioxide emission model of the UK housing stock. Ph.D. thesis, Leeds Metropolitan University, UK
62. Kadian R, Dahiya RP, Garg HP (2007) Energy-related emissions and mitigation opportunities from the household sector in Delhi. Energy Policy 35(12):6195–6211
63. Kashyap M, Khalfan M, Zainul-Abidin N (2003) A proposal for achieving sustainability in construction projects through concurrent engineering. In: Proceedings of the RICS foundation construction and building research conference. School of Engineering and the Built Environment, University of Wolverhampton, 1–2 Sept 2003
64. Kavgic M, Mavrogianni A, Mumovic D, Summerfield A, Stevanovic Z, Djurovic-Petrovic M (2010) A review of bottom-up building stock models for energy consumption in the residential sector. Build Environ 45:1683–1697

65. Kavousian A, Rajagopal R, Fischer M (2012) A method to analyse large data sets of residential electricity consumption to inform data-driven energy efficiency. Centre for integrated facility engineering (CIFE) working paper #WP130, Standford University, June
66. Keirstead J (2006) Evaluating the applicability of integrated domestic energy consumption frameworks in the UK. Energy Policy 34(17):3065–3077. https://doi.org/10.1016/j.enpol.2005.06.004
67. Kelly S (2011) Do homes that are more energy efficient consume less energy? A structural equation model for England's residential sector. EPRG working paper, electricity policy research group, University of Cambridge
68. Ko J, Fenner R (2008) Adoption of energy efficiency innovations in new UK housing. In: Proceedings of the institution of civil engineers energy, Nov 2008, pp 151–163
69. Kohler N, Hassler U (2002) The building stock as a research project. Build Res Inf 30 (4):226–236
70. Larsen BM, Nesbakken R (2004) Household electricity end-use consumption: results from econometric and engineering models. Energy Econ 26(2):179–200
71. Lee T, Yao R (2013) Incorporating technology buying behaviour into UK-based long term domestic stock energy models to provide improved policy analysis. Energy Policy 52:363–372
72. Levine M, Ürge-Vorsatz D, Blok K, Geng L, Harvey D, Lang S, Levermore G, Mongameli Mehlwana A, Mirasgedis S, Novikova A, Rilling J, Yoshino H (2007) Residential and commercial buildings. In: Metz B, Davidson OR, Bosch PR, Dave R, Meyer LA (eds) Climate change 2007: mitigation. Contribution of working group III to the fourth assessment report of the intergovernmental panel on climate change. Cambridge University Press, Cambridge, UK
73. Lins M, DaSilva A, Rosa L (2002) Regional variations in energy consumption of appliances: conditional demand analysis applied to Brazilian households. Ann Oper Res 117(1–4):235–246
74. Lutzenhiser L (1992) A cultural model of household energy consumption. Energy 17(1):47–60
75. Lutzenhiser L, Hackett B (1993) Social stratification and environmental degradation: understanding household CO_2 production. Soc Probl 40(1):50–73
76. Mcmanus A, Gaterell M, Coates L (2010) The potential of the code for sustainable homes to deliver genuine 'sustainable energy' in the UK social housing sector. Energy Policy 38 (4):2013–2019
77. Mhalas A, Kassem M, Crosbie T, Dawood N (2013) A visual energy performance assessment and decision support tool for dwellings. Visual Eng 1:7
78. Mihalakakou G, Santamouris M, Tsangrassoulis A (2002) On the energy consumption in residential buildings. Energy Build 34(7):727–736
79. Massachusetts Institute of Technology (MIT) (1997) Energy technology availability: review of longer term scenarios for development and deployment of climate-friendly technologies. Massachusetts Institute of Technology Energy Laboratory, Cambridge, MA, USA
80. Mohmed A, Ugursal VI, Beausoleil-Morrison I. (2008). A new methodology to predict the GHG emission reductionsdue to electricity savings in the residential sector. In: Proceedings of the third Canadian solar buildingsconference, Fredericton, 20th–22nd August.
81. Moll HC, Noorman KJ, Kok R, Engström R, Throne-Holst H, Clark C (2005) Pursuing more sustainable consumption by analyzing household metabolism in European countries and cities. J Ind Ecol 9:259–275
82. Motawa IA, Banfill PF (2010) Energy-efficient practices in housing—a system dynamics approach. In: Proceedings of 18th CIB world building congress: TG62—built environment complexity, Salford, UK, May, pp 44–56
83. Natarajan S, Levermore GJ (2007) Domestic futures–which way to a low carbon housing stock? Energy Policy 35(11):5728–5736
84. Natarajan S, Levermore GJ (2007) Predicting future UK housing stock and carbon emissions. Energy Policy 35(11):5719–5727

85. Natarajan S, Padget J, Elliott L (2011) Modelling UK domestic energy and carbon emissions: an agent-based approach. Energy Build 43:2602–2612
86. Oladokun MG, Odesola IA (2015) Household energy consumption and carbon emissions for sustainable cities—a critical review of modelling approaches. Int J Sustain Built Environ 4:231–247. Available at http://dx.doi.org/10.1016/j.ijsbe.2015.07.005
87. Palmer J, Cooper I (2012) United Kingdom housing energy fact file. DECC, London
88. Parekh A (2005) Development of archetypes of building characteristics libraries for simplified energy use evaluation of houses. In: Proceedings of the ninth international IBPSA conference, Montreal, Canada, pp 921–8
89. Parti M, Parti C (1980) The total and appliance-specific conditional demand for electricity in the household sector. Bell J Econ 11(1):309–321
90. Petersdorff C, Boermans T, Harnisch J (2006) Mitigation of CO_2 emissions from the EU-15 building stock. Environ Sci Pollut Res 13(5):350–358
91. Poortinga W, Steg L, Vlek C (2004) Values, environmental concern and environmental behavior: a study into household energy use. Environ Behav 36:70–93
92. Ruffell RJ (1977) An econometric analysis of the household demand for electricity in Great Britain. Scottish Academic Press, Scotland
93. Saidur R, Masjuki HH, Jamaluddin MY (2007) An application of energy and exergy analysis in residential sector of Malaysia. Energy Policy 35(2):1050–1063
94. Scottish Environmental Attitudes and Behaviour (SEAB) (2008) The Scottish environmental attitudes and behaviour survey 2008. Available at http://www.scotland.gov.uk/Publications/2009/03/05145056/11. Accessed 9 May 2012
95. Shipworth D (2005) Synergies and conflicts on the landscape of domestic energy consumption: beyond metaphor. Paper presented at the ECEEE summer study, Mandelieu La Napoule, France, pp 1381–1391
96. Shipworth D (2006) Qualitative modelling of sustainable energy scenarios: an extension of the Bon qualitative input–output model. Constr Manage Econ 24(7):695–703
97. Shorrock LD, Dunster JE (1997) The physically-based model BREHOMES and its use in deriving scenarios for the energy use and carbon dioxide emissions of the UK housing stock. Energy Policy 25(12):1027–1037
98. Shorrock LD, Henderson J, Utley JI (2005) Reducing carbon emissions from the UK housing stock. BRE report BR480. BRE Press, Watford
99. Stokes M, Rylatt M, Lomas KJ (2004) A simple model of domestic lighting demand. Energy Build 36:103–116
100. Stevenson F, Leaman A (2010) Evaluating housing performance in relation to human behaviour: new challenges. Build Res Inf 38(5):437–441
101. Stevenson F, Rijal HB (2010) Developing occupancy feedback from a prototype to improve housing production. Build Res Inf 38(5):549–563
102. Strachan N, Kannan R (2008) Hybrid modelling of long-term carbon reduction scenarios for the UK. Energy Econ 30:2947–2963
103. Summerfield AJ, Lowe RJ, Oreszczyn T (2010) Two models for benchmarking UK domestic delivered energy. Build Res Inf 38(1):12–24
104. Swan LG, Ugursal VI (2009) Modelling of end-use energy consumption in the residential sector: a review of modelling techniques. Renew Sustain Energy Rev 13:1819–1835
105. Swan L, Ugursal VI, Beausoleil-Morrison I (2008) A new hybrid end-use energy and emissions model of the Canadian housing stock. In: Proceedings of the COBEE conference, Dalian, China
106. Terry N (2011) Energy and carbon emissions: the way we live today. UIT, Cambridge
107. Tonn BE, White DL (1988) Residential electricity use, wood use, and indoor temperature: an econometric model. Energy Syst Policy 12(3):151–165
108. Tuladhar SD, Yuan M, Bernstein P, Montgomery WD, Smith A (2009) A top–down bottom–up modeling approach to climate change policy analysis. Energy Econ 31:S223–S234

109. Tweed C, Dixon D, Hinton E, Bickerstaff K (2014) Thermal comfort practices in the home and their impact on energy consumption. Arch Eng Des Manage 10(1–2):1–24
110. United Nations Environment Programme (UNEP) (2011) Buildings: investing in energy and resource efficiency. United Nations Environment Programme
111. Urge-Vorsatz D, Eyre N, Graham P, Harvey D, Hertwich E, Jiang Y, Kornevall C, Majumdar M, McMahon JE, Mirasgedis S, Murakami S, Novikova A (2012) Chapter 10—Energy end-use: building. Global energy assessment—toward a sustainable future. Cambridge University Press, The International Institute for Applied Systems Analysis, Cambridge, UK, pp 649–760
112. Van Raaij WF, Verhallen TM (1983) A behavioral model of residential energy use. J Econ Psychol 3(1):39–63
113. Wheelock J, Oughton E (1994) The household as a focus for comparative research. Centre for Rural Economy, Department of Agricultural Economics and Food Marketing, University of Newcastle-upon-Tyne, Newcastle
114. Williamson T, Soebarto V, Radford A (2010) Comfort and energy use in five Australian award-winning houses: regulated, measured and perceived. Build Res Inf 38(5):509–529
115. Wilson D, Swisher J (1993) Exploring the gap: top-down versus bottom-up analyses of the cost of mitigating global warming. Energy Policy 21(3):249–263
116. Yang J, Rivard H, Zmeureanu R (2005) Building energy prediction with adaptive artificial neural networks. In: Proceedings of the IBPSA ninth international conference, Montreal, Canada, 1401–1408
117. Yu Z, Fung BCM, Haghighat F, Yoshino H, Morofsky E (2011) A systematic procedure to study the influence of occupant behaviour on building energy consumption. Energy Build 43 (6):1409–1417
118. Yun GY, Steemers K (2011) Behavioural, physical and socio-economic factors in household cooling energy consumption. Appl Energy 88:2191–2200

Chapter 3
The Sociotechnical Systems of Energy and Carbon Emissions in Housing

3.1 Introduction

In Chap. 2 of this research book, it was concluded that there is a need for more robust approaches that have the capability of modelling the kind of complexity involved in housing energy and carbon emissions by taking into consideration the limitations of existing modelling approaches. It was argued in Chap. 1 of this book that the problem under investigation in this book sits squarely within the purview of STS. In this chapter, the systems idea that brought about the STS is first discussed by reviewing the rationale, historical background, as well as the basic concepts of the systems-based approach. Furthermore, the chapter discusses the theory of STS together with its basic concepts. The literature on the domain of the application and modelling techniques of STS is also reviewed accordingly. Additionally, the literature study specifically assesses and critiques the identified STS modelling techniques against a set of criteria that meet the requirements of the problem under investigation in this research book. Finally, the choice of the most suitable STS modelling technique concludes the chapter.

3.2 Systems Method of Scientific Inquiry

The systems idea of scientific inquiry did not come into being until the 1950s, when the main concepts and principles relating to the general systems theory were formulated. Banathy [3] noticed that the systems ideas of different fields share a common ground on systems orientations as those ideas embrace research or professional activities in the area of '*...system engineering, operations research, system dynamics, cybernetics and information science, general theory of systems, living systems and evolutionary theory, soft systems and critical systems theory, and chaos and complex systems theory*' [3]. As a result of this, these researchers now

M. G. Oladokun and C. O. Aigbavboa, *Simulation-Based Analysis of Energy and Carbon Emissions in the Housing Sector*, Green Energy and Technology,
https://doi.org/10.1007/978-3-319-75346-1_3

recognise the necessity of an interdisciplinary research field with the capability of coping with ever-increasing complexities that fall beyond the scope of a single discipline. The systems-based approach of scientific inquiry that emphasises the intrinsic order and interdependence of the complex problem in all its ramifications was therefore born.

The systems-based approach of scientific inquiry incorporates the systems theory, systems philosophy, and systems methodology as three main interrelated domains of disciplined scientific inquiry. While the systems theory and philosophy provide the philosophical basis for the systems-based approach, systems methodology gives the sets of methods, strategies, models, and tools for the systems-based approach of scientific inquiry. The rest of this section discusses the rationale behind the systems-based approach of scientific inquiry, the historical background of the systems-based approach, its concepts, components, and characteristics, as well as the types of systems.

3.2.1 Rationale Behind the Systems-Based Approach to Scientific Inquiry

Science is a way of acquiring testable knowledge about the world [17]. The classical method of scientific inquiry has played a prominent role in understanding and treating complexities in the 'world of science' and was adopted during the seventeenth and eighteenth centuries based on Descartes' analytic–deductive method which was used in studying complex phenomena. Clayton and Radcliffe [17] argue that science has a number of defining characteristics of which three are particularly important, namely 'replicability', 'refutability', and 'reductionism'. Descartes, however, bolstered reductionism by publishing Discourse on Method in 1637, and this publication gives four precepts which have influenced science for years [12]. According to Capra [12], these precepts are the following:

- Accept only that which you are certain of,
- Divide the topic into as small parts as possible,
- Solve simplest parts first, and
- Make as complete lists as possible.

This method breaks down the complex entities into small parts and studies them separately in order to gradually come to an understanding of the whole, which forms the philosophical basis of the classical view of scientific inquiry that gave rise to the technological and industrial revolutions in the globe [53].

Panagiotakopoulos [53] reports that by the end of the nineteenth century and during the twentieth-century complexity in the 'real world' expanded to such an extent that the classical method of scientific inquiry reached its limits in explaining the world. For this reason, Banathy [3] contends that the reductionist approach was no longer able to explain 'wholeness' which results from the mutual interaction of 'parts'. Premised on this, Skyttna [63] submits that to have a full understanding of the reason why a particular problem occurs and still persists, there must be an awareness of the parts in relation to the whole. This argument is, however, in direct opposition to the classical view of scientific reductionism and philosophical analysis as promulgated by Descartes [12]. In view of this, there is no doubt that simple tools cannot be used to capture ever-increasing complex problems in a world that is embedded in interconnected systems which operate in dynamically changing environments [3]. This therefore necessitates the needs for a shift from the classical approach paradigm to a systems approach of scientific inquiry.

The systems approach of scientific inquiry therefore represents a kind of paradigm shift which is now changing the emphasis from 'parts' to the study of 'whole' [4] since it is difficult to observe properties of the whole bit by bit. The systems approach of scientific inquiry therefore provides a multidimensional framework in which information from different disciplines and domains can be integrated without being forced into a one-dimensional mapping, which is not possible from the view of the classical approach.

3.2.2 Historical Background of the Systems-Based Approach

The account of the historical background of the systems-based approach of scientific inquiry has been widely reported in the literature. This subsection therefore gives a summary of how the approach has evolved over the years as reported by Decleris [20].

Banathy [4] reports the pioneer work of Ludwig von Bertalanffy in the area of biology in 1932 as being the first to develop the systemic idea on general systems theory in the early twentieth century. This work spurs research activities in many areas of scientific endeavours as it is capable of dealing with messy complex problems. Table 3.1 depicts how the systems approach has transcended over the years as well as the major actors during that time. Also, Table 3.2 shows how the ideas of the systems researchers have expanded owing to being confronted with more and more complex problems.

In Decleris [20], it was reported that the **first phase** of the systems-based approach evolution marked the start of the systems theory formulation between 1916 and 1940. This emanated as a result of the fact that the classical method of scientific inquiry found it difficult to cope with increasingly complex problems and the need arose to depart from the thinking of traditional analytic–deductive approach. This is where the basic systemic ideas are then defined and announced as depicted in Table 3.2. The **second phase** of this evolution is termed the *practical orientation* phase as the period witnessed a more practical relevance of the technological revolution. This is the period when the technological advancement skyrocketed mainly because of the need to tackle many complex problems that arose

Table 3.1 Systems approach evolution

Systems approach evolution	Main contributors
First phase (1916–1940): *Precursors*	• **Biology:** Bertalanffy (1932) • **Linguistics:** Saussure (1916) • **Psychology:** Kohler (1929) • **Anthropology:** Malinowsky (1926), Radecliffe-Brown (1935) • **Sociology:** Parsons (1937)
Second phase (1940–1945): *Practical orientation*	• **Operations research:** Rowe (1941), Rand (1946) • **Policy science:** Lasswell (1951)
Third phase (1945-1970): *Formation —'Hard' systems*	• **Mathematical theory of communication:** Shannon (1948) • **Cybernetics:** Wiener (1948), Ashby (1952) • **Network and linear systems theory:** Guillemin (1957) • **General systems theory:** Bertalanffy (1956) • **Society for general systems research:** Boulding (1954) • **Systems analysis:** Riemann (1953) • **Systems construction:** Riemann (1953)
Fourth phase (1970–1992): *Spreading—'Soft' systems*	• **Systems management:** Churchman (1979) • **Social systems construction:** Pinch et al. (1984) • **Human systems:** Rosenstiel et al. (1991) • **National systems organisation—International institutes:** Trist (1981)
Fifth phase (1992–): *'Hard' and 'Soft' systems*	• **Climate change:** IPCC (2010) • **Sustainability:** Redclift (2005)

Adapted with extension from Decleris [20]

Table 3.2 Systemic ideas evolution

Systems approach evolution phases	Problem	Idea
First phase (1916–1940)	Description of complex objects/ phenomena	Definition and announcement of basic systemic ideas
Second phase (1940–1945)	Coordination of complex actions	Compilation of optimum theoretical model (modelling/optimisation)
Third phase (1945–1970)	Information transmission, construction of complex artefacts	Information (information and control systems)
Fourth phase (1970–1992)	Uncertainty and change in complex systems behaviour	Dynamic systems, information processing systems
Fifth phase (1992–)	Global climate change	Sustainability

Adapted with extension from Decleris [20]

from World War II. The main complex problem that developed then was that of the coordination of complex actions (Table 3.2), which gave rise to the main idea of compilation of optimum theoretical models based on the operational research and policy science disciplines (Tables 3.1 and 3.2).

The **third phase** witnessed the development of a highly sophisticated systems-based technological advancement that contributed in no small measure to the birth of new scientific disciplines. This causes the systems approach to be reoriented and refined in order to accommodate these developments. The phase saw the emergence of the mathematical theory of communication, cybernetics, network and linear systems theory, systems analysis, and systems construction. In fact, the phase is known for the formation of *hard* systems (Table 3.1). The major problem of this phase is in the area of information transmission and the construction of complex artefacts, which brought about the ideas of information and control systems (Table 3.2). Decleris [20] argues that subsequent inventions in the fields of cybernetics and mathematical theory of communication had a remarkable impact on the world of science today, especially in the area of human–machine systems (i.e. STS).

The **fourth phase** is regarded as the formation of *soft* systems. This is because the period saw an enormous expansion in the systems approach across many fields of study in a bid to solve complex social problems. Principally, problems in human systems are solved by this approach. During this period, the problem of uncertainty and change in complex systems behaviour grew and became difficult to manage owing to the *soft* nature of the problem. The main ideas being communicated here to solve the problem were the ideas of dynamic systems and information processing systems (Table 3.2). These ideas were seen to be effective in disciplines such as systems management, spatial planning, education, and transportation, among others.

Arguably, the world is in the **fifth phase** of systems approach where complex problems due to global climate change have surfaced in the last two decades involving *soft* and *hard* systems. In order to solve the problem of climate change, the main idea here is sustainability, especially regarding issues of carbon emissions, under the purview of which the study in this research book falls. This problem then necessitated the use of a systems approach to capture the problem.

Banathy [4] regards researchers such as Ashby, Bertalanffy, Boulding, Fagen, Gerard, and Rappoport as the pioneers that set forth the basic concepts and principles of the general theory of systems that metamorphosed into the systems-based approach today. The concept of a systems approach advocates that the properties and characteristics of the *whole*, which is the *systems* itself, are quite different from summing up the *parts* in such a way that properties of a *whole* cannot be observed bit by bit as against the view of classical, traditional method of scientific inquiry that studies *parts* with linear *cause and effect*. Banathy [3] argues that deterministic linear *cause and effect* is practically inadequate in dealing with many interactive variables of complex, dynamic systems. In contrast, the systems-based approach is able to capture the dynamics of multiple, mutual, and recursive complex causation

[3] and sees the behaviour of the systems as nonlinear, nondeterministic, and expansionist in nature, negating the reductionist approach of classical science. Based on the ideas of the pioneers of the systems-based approach, Banathy [3] proposes an overview of the key distinctions between the classical view of scientific

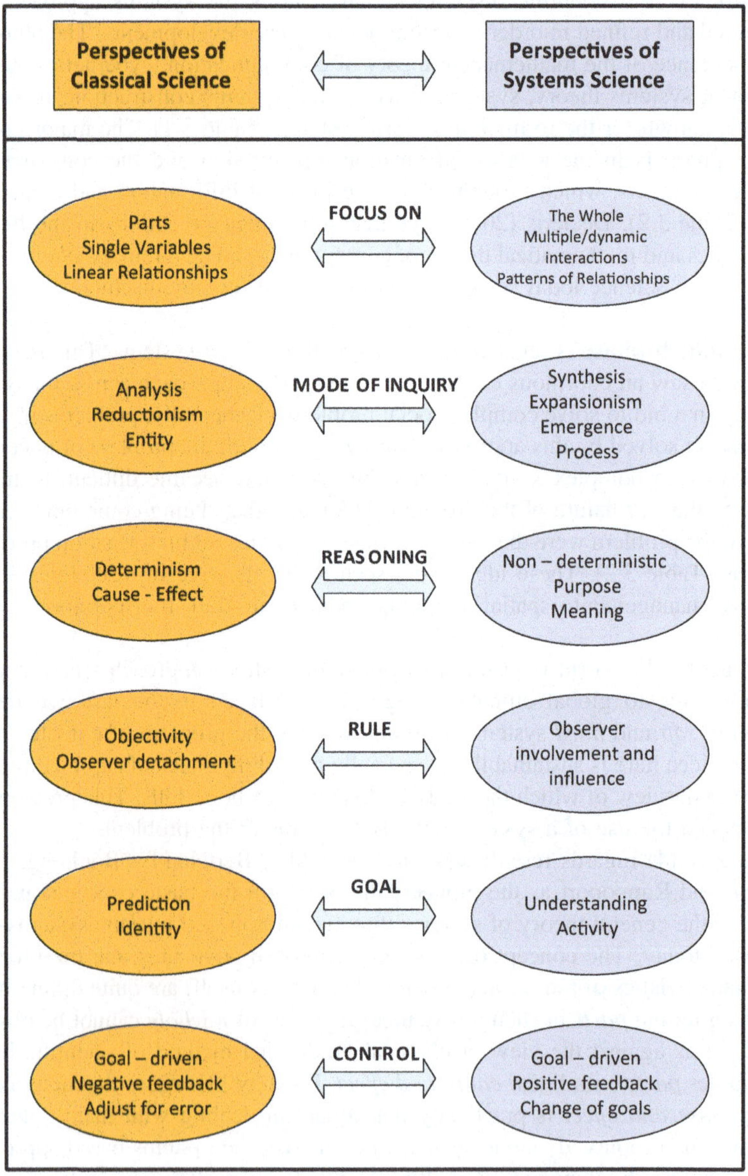

Fig. 3.1 Key distinctions between classical and systemic orientations. Adapted from Banathy [3]

inquiry and the systems view of scientific inquiry. These distinctions are based on what they 'focus on', their 'mode of inquiry', the way they 'reason', the 'rule' guiding them, their 'goal' and finally 'control' as shown in Fig. 3.1.

3.2.3 *Components and Characteristics of Systems-Based Approach*

Based on the basic concepts of the systems-based approach explained above and the work of researchers such as Turner [70], Capra [12], and Blockley [6], it is evident that the systems-based approach encapsulates many interrelated components, the properties of which are altered should the systems be cloaked in any way [74]. These components are the main anchor of the systems-based approach. To this end, adopting a systems approach to solving the problem relating to HECE in the context of this research book entails gaining insights into the effects of interactions among different variables hypothesised to influence household energy consumption.

By drawing from the studies of Turner [70] and Blockley [6], Waterson [74] was able to capture those components and their characteristics that are the central idea being communicated by the systems-based approach as shown in Fig. 3.2. The characteristics are therefore similar to the problem under investigation in this book. The three main components given by Waterson [74] (Fig. 3.2) are the following:

- Input–output processes: This aspect gives the relationships that exist between the systems inputs and their corresponding outputs containing elements such as multifinality, equifinality. Whole–part relationships: The main idea being communicated by this component is hinged on the fact that the working of the systems as a whole needs firstly to be analysed in parts, as suggested by Gibson [27]. The component further suggests that the whole is more than just summing up the parts [3] as the kind of relationships existing between them is argued to be complex, dynamic, and chaotic in nature [62]. Holism, entropy, and system elements are therefore expressed as the major elements of this component.
- Connectivity between elements: This component expresses the interrelationships among different elements within the systems in terms of hierarchy, interactions, and regulation. The complexity of the systems here is therefore elaborated based on the causal relationships and feedback structure(s?) among these elements [38].

Additionally, Wilson et al. [76] and Walker et al. [73] offer the description of behaviour exhibited by the system in order to further capture its characteristics. They argue that a system as a dynamic and complex whole containing integrated interacting functional parts has energy, material, and information flowing through it. These energy, material, and information of the studied systems are placed within

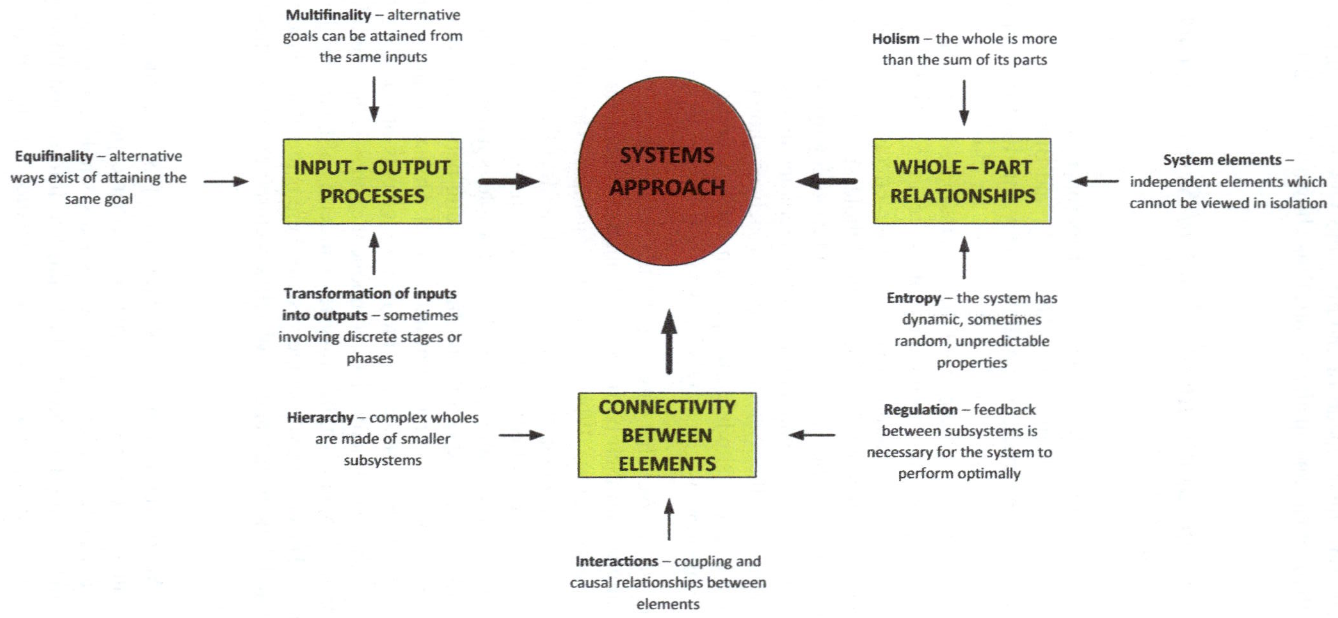

Fig. 3.2 Components and characteristics of the systems approach. Adapted from Waterson [74]

an environment surrounded by permeable boundaries, which are capable of exhibiting erratic behaviour while its elements seek equilibrium.

3.2.4 Types of Systems

The study of Decleris [20] classified systems into *hard* and *soft* systems as evidenced from Table 3.1. *Hard* systems, for example, are described as technical and physical systems that can be quantified, while their behaviour can be fully controlled at the same time [53]. However, these cannot easily take unquantifiable variables into consideration. Different from the *hard* systems, *soft* systems are good at capturing and understanding unquantifiable variables such as people's opinions, cultures, viewpoints. In short, they will address qualitative aspects of any problem situations. To this end, the classification brought about the concept of STS as a systems-based approach capable of handling the complexity posed by the interaction of 'human' and 'machine', which is good at combining the quantitative and qualitative research strategies. The next section then discusses the STS theory.

3.3 The Sociotechnical Systems Theory

STS theory has evolved over the years as a kind of coaction among the sociologists that specialised in a new area of academic endeavour termed the 'sociologists of technology' [21]. There was a general belief that engineers and technologists, for example, tend to ignore the importance of the socio-aspect of their work, while on the other hand, the social scientists tend not to know much about the technology and are therefore reluctant to consider the artificial reality of technical objects [58]. STS theory has then been used as the theory that combines the two divides together. Therefore, the STS theory serves as the theoretical basis for this study. The rest of this section discusses the basic concepts of the STS theory and its domain of application.

3.3.1 Basic Concepts of Sociotechnical Systems Theory

The origin of the concepts of STS as a methodology for the systems-based approach of scientific inquiry can be traced to the studies undertaken by the Tavistock Institute, in London, especially during the post-war reconstruction of industry [16]. Cartelli [16] reports that the emergence of the concepts was necessary in pursuit of a fit between the workforce and machine during the introduction of technological systems for work automation when it was found that workers are resistant to technological innovation. Since then, the concept has come into prominence and serves as the theoretical framework underpinning many studies.

According to Walker et al. [73], STS as a concept is founded on two main principles. The first one is the interaction between the social and technical subsystems that set the conditions for successful (or unsuccessful) systems performance. They argued that the interactions are comprised partly of linear '*cause and effect*' relationships, the relationships that are normally '*designed*', and partly from '*nonlinear*', complex, and even unpredictable relationships, which are those that are often unexpected. *Soft*, which is socio, does not necessarily behave like the *hard*, which is technical [73]. Additionally, Walker et al. [73] contend that the growth in complexity and interdependence makes the 'technical' systems, for example, start to exhibit nonlinear behaviours. As such, the STS as a technique of the systems-based approach of scientific inquiry is used to handle this kind of complexity as both the methodology and tools. The second of the two main principles is founded on '*joint optimisation*' of the two systems.

Interestingly, Dwyer [21] illustrates the concept of STS by the use of a generic model as shown in Fig. 3.3. According to her, STS is seen to contain *components* that are referred to as *social structures* and *artefacts* that are called *technical elements*, which contribute directly or through other *components* to a common *system goal*. It was shown that both the *components* and *artefacts* interact with each other. What is guiding the overall behaviour of the system is the system goal.

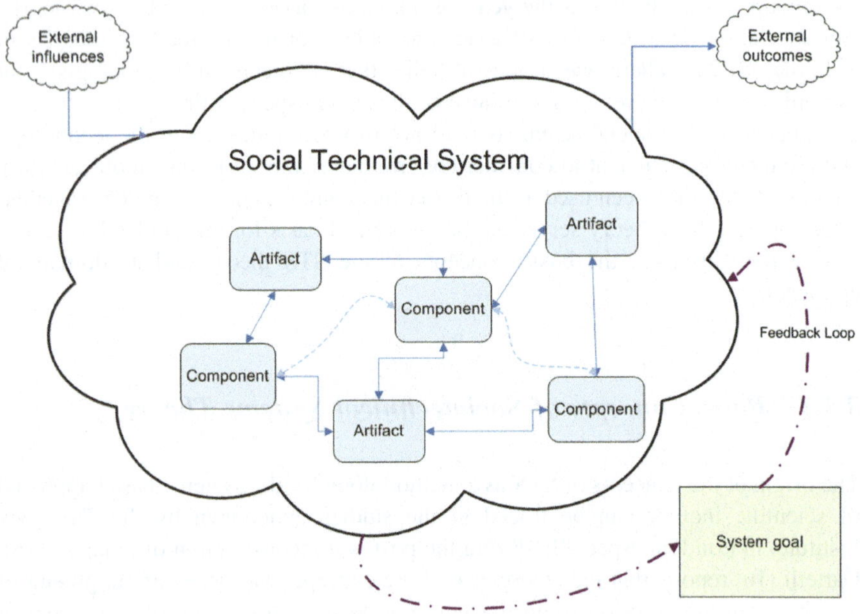

Fig. 3.3 A model of a sociotechnical system. Adapted from Dwyer [21]

The feedback loop enables the actual output of the system to be compared to the system goal. Hughes [31] argues that it is only through the feedback loop that errors are detected and thereby corrected in order to have an improvement in system performance.

The central idea from the concepts of the STS can be applied to the issue of HECCE in order to contextualise this discussion. Household systems consist of an interplay among the dwelling systems (in terms of dwellings' physical characteristics and technological systems put in place within them) referred to as technical systems (*artefacts*), occupants' systems (in terms of behaviour towards energy consumption, for example) referred to as sociosystems (*components*), and external environment systems (in terms of external temperature, energy prices, and the like) referred to as technical or sociosystems, or both. These systems are then interrelated and influence energy consumption and associated carbon emissions.

A detailed analysis of all the variables in the systems suggests that they all depend on one another, thereby making the systems complex. This is mainly so because the variables within each of the systems have multiple interdependencies with multicausal feedback structures considering their effect on energy consumption and carbon emissions. Further, they are interconnected, chaotic in nature, and difficult to understand, predict, and keep under control, thereby calling for a pragmatic approach such as the STS approach to handle the situation under consideration. An appraisal of the problem in this research book suggests that the STS approach is adequate in capturing it. Hence, STS theory serves as the theoretical background to the research.

3.3.2 Domain of Application of Sociotechnical Systems

A review of the domain of application of STS by various researchers was undertaken. During the review, it was noted that the concept of STS has slightly different connotations to researchers in different fields of study; for example, in engineering, it means that organisational form follows technical function, while technical function too follows organisational form; in computer science, the technical system consists of hardware and software that make an information system, while the users of this system and the organisation in which it is embedded form the social system. However, similarities in the use of this concept are stronger and more numerous than the differences. A literature search was then conducted irrespective of the definition used for the STS by different researchers.

Based on the review, STS has been successfully implemented in human–computer interaction studies, information technology, software engineering, engineering (general), business and management, medicine, and a host of others. For example, De Greene [29] used STS in the context of organisational design management. Likewise, Appelbaum [1] used STS in the context of organisational development where it was argued that the integration of organisational development with technological advancement into a total system could prove difficult, but the use of STS would make it possible. Also, STS was used in the context of innovation which predisposes systemic changes in any organisation [26]. Williams and Edge [75], Rohracher [57] and Geels [25] used STS in the context of diffusion of technology in an organisation.

Further, STS has been used in energy supply and demand, especially when it was necessary to study the sociotechnical influences on energy use [49, 61, 60]. STS has been used in computer/software engineering as well as in communication and telecommunication engineering [54]. This concept of STS has also found application in the domain of water management when considering an irrigation project [34] and in the domain of agriculture and food [44]. The above then shows how the range of research has been extended using STS to solve real-life problems.

STS has been previously used as methodology to model the complexity of real systems' elements and relationships as indicated above. STS is difficult to model because of its complex nature. It is complex because its elements have multiple interdependencies and have a multicausal correlation structure. Further, STS exhibits a kind of nonlinear behaviour where changes in input are neither proportional to changes in output, nor is the input-to-output relationship fixed over time [49]. The ability of STS to integrate both 'hard' and 'soft' data together under the conditions described above makes it different from other complex systems.

3.4 Methods of Modelling the Sociotechnical Systems

This section covers the techniques utilised by different researchers to model STS under different themes. Based on the review conducted in Sect. 3.3.2, a detailed analysis of selected articles from the pool of articles reviewed was undertaken. Specifically, these articles were analysed for the modelling techniques utilised in the context of STS. The articles were then analysed according to the STS domain, STS definition, whether modelling or simulation was performed, the modelling or simulation techniques that were utilised, whether the results produced are reproducible, whether the techniques presented are capable of being generalised to other domains of application, and whether the model can be easily extended and if it can be, to what extent this can be done. The main aim of this exercise is to identify the major techniques that have been used by different studies to conceptualise STS problems. Table 3.3 shows the results of this review. As shown in Table 3.3, the articles reviewed were assessed to indicate the presence of any evidence to suggest within their body whether there is a match or no match or unclear in a STS application domain, the STS definition, modelling/simulation, modelling/simulation technique, and whether it is reproducible, generalisable, and extendable. The (+) sign indicates that there is a match, whereas a (−) sign shows that there is no match. The (?) sign signifies that evidence of those criteria is unclear.

The results of the review conducted show that most of the articles analysed explicitly indicate STS as the domain of application for their studies. Also, about half of those articles claim that the STS method presented can be generalised.

Table 3.3 Review of modelling techniques for sociotechnical systems

Article authors	STS application domain	STS definition	Modelling/ simulation	Modelling/simulation technique	Reproducible	Generalisable	Extendable
Bergman et al. [5]	+	+	+	ABM	?	+	?
Carley [13]	+	+	–	?	–	+	–
Iivari and Hirschheim [32]	–	–	–	?	–	+	–
Jarman and Kouzmin [33]	–	–	–	?	–	+	–
Shipworth [61]	+	+	–	–	+	?	?
Sterman [65]	+	–	+	SD	+	+	+
Olla et al. [52]	+	–	+	ANT	–	?	–
Cai et al. [8]	+	–	+	FL	+	?	–
Li et al. [41]	+	–	+	FL	+	–	–
McNeese et al. [47]	–	+	–	–	–	+	–
Ramanna et al. [55]	–	–	–	–	–	?	–
Shah and Pritchett [59]	–	+	+	ABM	–	+	+
Johnson [37]	–	+	–	?	–	+	–
Shipworth [60]	+	+	+	BBN	+	+	+
Smajgl et al. [64]	?	?	–	–	–	–	–
Masys [45]	+	?	–	?	–	+	–
Sterman [67]	+	–	+	SD	+	+	+
Sutcliffe et al. [68]	–	+	–	–	–	?	–
Thissen and Herder [69]	+	+	–	–	–	+	–

(continued)

Table 3.3 (continued)

Article authors	STS application domain	STS definition	Modelling/ simulation	Modelling/simulation technique	Reproducible	Generalisable	Extendable
Yahja and Carley [78]	–	–	+	ABM	?	+	?
Lock [43]	+	–	+	CM	–	?	–
Ritchey [56]	+	–	+	MA	–	?	–
McIntosh et al. [46]	+	+	–	?	–	?	–
Lock [42]	+	–	+	CM	–	?	–
Wu and Xu [77]	+	–	+	SD and FL	+	+	+
De Waal and Ritchey [72]	+	–	+	MA	–	?	–
Jensen [36]	+	–	+	BBN	+	+	+
Callon [10]	+	–	+	ANT	–	?	–
Natarajan et al. [50]	+	+	+	ABM	+	–	+
Feng et al. [22]	+	–	+	SD	+	+	+
Carroll et al. [14]	+	+	+	SNA	–	–	–
Carroll [15]	+	+	+	ANT and SNA	+	+	–

ANT Actor–network theory, *ABM* Agent-based modelling, *BBN* Bayesian belief network, *CM* Configuration modelling, *FL* Fuzzy logic, *MA* Morphological analysis, *SNA* Social network analysis, *SD* System dynamics. '+' means there is a match, '–' means there is no match, and '?' means unclear

Furthermore, the analysis shows that just some of the STS approach presented can be reproduced and further extended to accommodate additional modules or subsystems. It was also concluded from the review that out of 32 articles analysed, 20 of them provided the modelling/simulation techniques utilised for their different studies within the context of STS. Therefore, the output of the study shows some of the techniques that have served as decision support tools or platforms under which the STS of real problems is modelled. In this regard, this study therefore identified the following as techniques for modelling STS:

1. Actor–network theory (ANT)
2. Agent-based modelling technique (ABM)
3. Bayesian belief network (BBN)
4. Configuration modelling (CM)
5. Fuzzy logic (FL)
6. Morphological analysis (MA)
7. Social network analysis (SNA)
8. System dynamics (SD).

The next section therefore summarises and critiques these modelling techniques.

3.5 Critique of the Methods

For any of those techniques to be adequate in the context of this research book, there are some criteria they must fulfil based on the nature of the problem under investigation in this book. For example, different researchers have criticised the energy models in the housing sector for their lack of transparency [39, 48] as discussed in Chap. 2. Also, Hitchcock [30], Kohler and Hassler [40], and Shipworth [61, 60] established that the complex sociotechnical systems are highly interdependent, chaotic, and nonlinear, and problems involving these are better solved using a pluralistic approach.

Therefore, it is important to set the criteria according to which the STS modelling techniques will be compared. And as such, the modelling techniques are compared to one another based on (1) transparency, (2) multiple interdependencies, (3) dynamic situations, (4) feedback processes, (5) nonlinear relationships, (6) *hard* and *soft* data, (7) uncertainties of the variables involved, (8) chaotic assumptions, and (9) the use of the model as a learning laboratory. Against this background, the techniques were all assessed, compared, and critiqued in order to decide which one of them will be able to capture the problem under investigation based on the above criteria. Table 3.4 summarises and compares all the STS modelling techniques. The tenets as well as strengths and weaknesses of each of the STS modelling techniques are therefore discussed accordingly in the following subsections. This exercise would, undoubtedly, help in identifying which of them is best for conceptualising the problem under investigation in this research book.

Table 3.4 Comparative analysis of STS modelling techniques

Criteria	ANT	ABM	BBN	CM	FL	MA	SNA	SD
Transparency	√	√	√				√	√
Multiple interdependencies	√	√	√				√	√
Dynamic situations		√						√
Feedback processes								√
Nonlinear relationships	√	√	√	√	√	√	√	√
Considering 'hard' and 'soft' data	√	√	√	√	√	√	√	√
Chaotic assumptions	√	√	√	√	√	√		√
Uncertainties		√	√	√	√	√		√
Learning laboratory tool			√					√

3.5.1 *Actor–Network Theory*

Actor–network theory (ANT) was first proposed by Michel Callon and Bruno Latour [10, 11]. Olla et al. [52] argue that ANT provides a platform for understanding the creation of networks of aligned interests where, according to Olla et al, [52], the world is full of hybrid entities containing both human and nonhuman elements. Carroll [15] contends that the greatest strength of ANT lies in its ability to integrate both *hard* and *soft* data (Table 3.4). Also, the approach is capable of modelling problems containing variables that have multiple interdependencies with nonlinear relationships under chaotic assumptions. It, therefore, has some merit in modelling STS problems. However, the approach has been criticised for its inability to provide the means of differentiating between human and nonhuman elements within the model [15].

3.5.2 *Agent-Based Modelling*

According to Jennings [35], an agent is seen to be an entity or component that is autonomous, reactive, proactive and capable of social interaction. Agent-based modelling (ABM) aims to model the global consequences of each of the entities or components of a system, including their behaviour and interactions. This is then the main distinguishing element that sets agent-based models apart from other models [19]. In general, the ABM approach is applicable for the modelling of complex systems if the following conditions are satisfied [71]:

- The problem has a distributed character;
- The subsystems operate in a highly dynamic environment;
- The subsystems have to interact in a flexible way; and

- The subsystems are characterised by reactivity, proactiveness, cooperativeness, and social ability.

As shown in Table 3.4, ABM seems to be a suitable approach to create models of STS because of its capability to handle both *hard* and *soft* data with multiple interdependencies and to treat nonlinear behaviour of such data set under small uncertainties [5, 50]. To this end, a number of studies have utilised the approach for modelling complex problems. For example, the study of Yahja and Carley [78] used the approach to model improvements in multiagent social network systems. Also, Natarajan et al. [50] found the approach useful in modelling energy consumption and carbon emissions of the UK housing stock. However, the approach has some drawbacks. For example, its weakness lies in its inability to handle multiple feedback processes and its difficulty in being used as a learning laboratory.

3.5.3 Bayesian Belief Networks

Bayesian belief networks (BBNs) were developed around the late 1980s, and their applicability did not come into prominence until the 1990s. According to Jensen [36], BBNs emerged as an intuitive technique for reasoning under uncertainty. This technique combines different data types as well as learning from new observations as they become available. Advantages (Table 3.4) of using BBNs as opined by Gill [28] are the following:

- The ability to learn as new information is received or population variables change;
- The capacity to systematically integrate a wider variety of data types and any prior available knowledge;
- Predictions about the likely future state of the system are allowed based on what is currently known about the system and assumptions about future data;
- The capability to learn causal relationships and gain understanding of a problem domain and then predict the consequences of intervention;
- Overt and clear model assumptions; and
- Straightforward sensitivity testing.

This approach has been successfully used in a number of applications. Applications of BBN in the field of environmental management include management of fisheries [23], land use change [2], agricultural land management [9], and integrated water resource management [7]. This approach has also been applied to modelling the sociotechnical influences on domestic energy consumption in one of the UK's Carbon Vision programmes, namely the Carbon Reduction in Building (CaRB) project [61, 60]. As argued by Shipworth [61], however, BBN is used as decision support systems mainly because of their capability to integrate different arrays of data, as well as synthesising relevant factors in social, economic, ecological, and technical fields which then make it particularly useful in the complex socio-economic and sociotechnical environments of sustainable development. However, the BBN approach is not without its own drawbacks. De Waal and Ritchey [72] argue that using BBN may prove a little difficult during the initial problem formulation phase of the modelling process and difficult to deal with time-dependent data set with feedback processes.

3.5.4 *Configuration Modelling*

Configuration modelling (CM) is another technique and decision support tool recently proposed by Simon Lock [42, 43] for modelling the STS. Lock [43] acknowledges that managing the evolution of large systems is a complex and difficult task where the full social and technical implications of any proposed changes must be fully appreciated before a decision is made whether to proceed with their implementation. He contends that the task becomes challenging and difficult to manage since the interplay between the technical and nontechnical components is often complex and the various human factors that are involved inject much variability and unpredictability into the system. It is against this backdrop that a new decision support tool that permits the investigation and exploration of different configurations of socio and technical components needed in order to fully predict how changes made to the individual components or the overall configuration of a system will affect the operational behaviour of that system during the real-world operation [43]. Lock [42], however, argues that this modelling paradigm is a novel approach in the sense that it is easy and quick to construct and, in addition, can help to promote understanding of different stakeholders. However, there is a lack of evidence from the body of literature to suggest that this approach has the capability to capture multiple interdependencies of data sets under dynamic conditions. Furthermore, the domain of application of this approach has been limited to the area of software engineering as this has not gained a wider application, but has some merits in modelling STS.

3.5.5 *Fuzzy Logic*

The capability of fuzzy logic (FL) to model STS has been highlighted in literature. FL began with the 1965 proposal of fuzzy set theory by Lotfi Zadeh [79]. It is a mathematical approach that is used to represent uncertain and imprecise information. Cai et al. [8] argue that this method deals with reasoning that is approximate rather than fixed and exact, but is effective in describing highly complex, ill-defined mathematical systems. Furthermore, the approach can effectively support linguistic imprecision and vagueness [41]. A number of studies have used this approach to model complex systems under different themes. For example, Cai et al. [8] used the approach to identify optimal strategies within energy sector planning under multiple uncertainties of variables involved. Also, under the same theme as Cai et al. [8], Li et al. [41] combined the approach with stochastic programming to model energy and environmental planning systems. Further to this, Wu and Xu [77] combined FL with SD to predict and optimise the energy consumption of World Heritage areas in the People's Republic of China. While the major strength of the approach lies in its ability to model systems under varying degrees of uncertainty, it does have some limitations that may prevent it from being used within the context of this research book. It lacks the ability to handle multiple interdependencies of variables under dynamic situations. Also, it does not support feedback processes and cannot be used as a learning laboratory. As can be seen, the strengths and limitations of this technique are profound as succinctly summarised in Table 3.4.

3.5.6 Morphological Analysis

Morphological analysis (MA) was developed by Zwicky, a Swiss-American astrophysicist and aerospace scientist, as a general method for structuring and investigating the total set of relationships contained in multidimensional, usually nonquantifiable complex problems [56, 80]. The concept and application of MA as strategic decision support are closely related to BBNs. According to De Waal and Ritchey [72], it allows small groups of subject specialists to define, link, and internally evaluate the parameters of complex problem spaces easily, thus creating a solution space and flexible inference. However, they argued that MA cannot easily treat hierarchal structure and causal relationships, but when combined with BBNs, the benefits of both of these techniques can be optimised. This technique has previously been applied to diverse fields based on the work of Zwicky. Among these are astrophysics, the development of propulsive power plants and propellants, and the legal aspects of space travel and colonisation [72]. The suitability of this approach to the area of application of this research book is limited, though it has some potential when combined with other suitable approaches as shown earlier in Table 3.4.

3.5.7 Social Network Analysis

Social network analysis (SNA) views social relationships in terms of network theory that consists of *nodes* and *ties* (also called *edges*, *links*, or *connections*). Nodes, according to [24], are individual actors within the networks, and ties are the relationships between the actors. The resulting graph-based structures are often very complex. There can be many kinds of ties between the nodes. Research in a number of academic fields has shown that social networks operate on many levels, from families up to the level of nations, and play a critical role in determining the way problems are solved, organisations are run, and the degree to which individuals succeed in achieving their goals. Most importantly, SNA has the capability of modelling nonlinear, multiple interdependent quantitative and qualitative variables [14]. Therefore, it has some merits in modelling STS, but its strength could be improved upon when combined with other approaches.

3.5.8 System Dynamics

System dynamics (SD) emerged in the 1950s when introduced by Jay Forrester as a multidisciplinary field of study that has the capability to deal with complex systems. SD, as a systems-based approach, is regarded as a methodological approach and set of analytical tools for modelling STS [49]. Ogunlana et al. [51] mention that the SD is a useful approach for managing processes with two major characteristics:

- They involve changes over time, and
- They allow feedback, and the transmission and receipt of information.

Interestingly, Coyle [18] offers a robust definition of SD as a method '*…that deals with the time-dependent behaviour of managed systems with the aim of describing the system and understanding through a qualitative and quantitative model, how information feedback governs its behaviour, and designing robust information feedback structures and control policies through simulation and optimisation*'.

Over the years, the approach has developed itself into a very powerful tool for modelling complex systems. As a result, it has found wider application in an array of different fields. For example, Ogunlana et al. [51] used it in the field of project management and Feng et al. [22] in the area of energy consumption and carbon emissions, and a host of other applications. The approach was able to garner use in different capacities based on its strength. Accordingly, Sterman [66] justifies the application of SD to modelling complex problems in the sense that:

- SD models are well suited to capturing multiple interdependencies;
- SD was developed to deal with dynamics;
- SD is the modelling method of choice where there are significant feedback processes;
- SD, more than any other modelling technique, stresses the importance of nonlinearities in model formulation and therefore is able to capture any form of nonlinear relationships; and
- SD modelling permits both 'hard' and 'soft' data.

However, SD has a limited capability of handling situations under uncertainties. This weakness has received due attention from the SD research circles, and significant improvements have been made on this as some of the SD software now incorporates optimisation and sensitivity analysis of uncertain parameters.

3.5.9 Conclusion from the STS Modelling Techniques for the Research

The above summary and critique of different STS modelling techniques give the appropriateness of each of the techniques in conceptualising the problem under investigation in this book. Of the nine criteria used in appraising the techniques, the analysis suggests that SD meets almost all the nine criteria, except for its inability to fully handle parameters under uncertainties, of which a full-scale improvement on this aspect is underway. As mentioned in Sect. 3.5.8, the SD method was specifically introduced by Jay Forrester in order to handle complex problems that have multiple interdependencies and are dynamic in nature with many feedback structures. The tools for this technique have in-built functions to capture the nonlinear relationships existing among different variables making up the model with the capability of accepting both qualitative and quantitative data and converting the same to simulation. The technique can also handle chaotic situations by invoking the delay functions built into the tools. It is necessary to mention that the technique is undergoing a constant review, and over the years, its transparency aspect has been greatly enhanced and improved upon. This means that all the model variables, including the algorithms, can be assessed and scrutinised by third parties. In summing up all these characteristics of SD, it appears more appropriate to conceptualise the problem under investigation in the context of this research book.

However, there are other techniques that meet substantial numbers of the criteria of assessment of the techniques. For example, both ABM and BBN each meet seven of those criteria. In ABM, the models developed using the technique can be easily scrutinised for their algorithms. The major drawback is in its inability to handle feedback processes which has been argued as germane to the dynamic characteristic of any of the techniques. Also, the approach cannot be used as a learning laboratory where policies can be tested for results of implementation before actually being implemented in reality. In the case of BBN, the technique is transparent as well. Clearly, it is unsuitable for this research book because of its inability to handle dynamic situations involving feedback processes.

3.6 Summary

The main aim of this chapter was to identify the most suitable modelling approach to conceptualise the complex sociotechnical systems of energy consumption and carbon emissions in the housing sector. The chapter identified this modelling approach. Before the review of literature for modelling techniques of STS, the chapter reviewed the literature on the systems-based approach of scientific inquiry as it forms the theoretical knowledge base underpinning the STS. This is mainly to give the philosophical background of STS. A literature search was then conducted, and the review results revealed that the domain of application of STS has been in the areas of human–computer interaction studies, information technology, software engineering, engineering (general), business and management, medicine, and a host of others. Also, the review was analysed for modelling techniques for STS. The following techniques were identified, namely actor–network theory, agent-based modelling technique, Bayesian belief network, configuration modelling, fuzzy logic, morphological analysis, social network analysis, and system dynamics. These techniques were further probed for their capability in capturing the problem under investigation in this book against a set of criteria. After a careful appraisal of all the techniques, the study identified system dynamics as the most suitable technique in conceptualising the problem under investigation in the context of this research book.

References

1. Appelbaum SH (1997) Socio-technical systems theory: an intervention strategy for organizational development. Manag Decis 35(5/6):452
2. Bacon PJ, Cain JD, Howard DC (2002) Belief network models of land manager decisions and land use change. J Environ Manage 65(1):1–23
3. Banathy B (2000a) A taste of systemics. In: The first international electronic seminar on wholeness of a special integration group of the international society for the systems sciences. Available at: http://www.newciv.org/ISSS_Primer/asem04bb.html
4. Banathy B (2000b) The evolution of systems inquiry. In: The first international electronic seminar on wholeness of a special integration group of the international society for the systems sciences. Available at: http://www.newciv.org/ISSS_Primer/seminara.html
5. Bergman N, Haxeltine A, Whitmarsh L, Köhler J, Schilperoord M, Rotmans J (2008) Modelling socio-technical transition patterns and pathways. J Artif Soc Soc Simul 11(3):7
6. Blockley D (1998) Managing proneness to failure. J Conting Crisis Manage 6(2):76–79
7. Bromley J, Jackson NA, Clymer OJ, Giacomello AM, Jensen FV (2004) The use of Hugin to develop Bayesian networks as an aid to integrated water resource planning. A report on the MERIT (Management of the Environment and Resources using Integrated Techniques) EU Framework V funded project, Centre for Ecology and Hydrology, Wallingford
8. Cai YP, Huang GH, Yang ZF, Tan Q (2009) Identification of optimal strategies for energy management systems planning under multiple uncertainties. Appl Energy 86(4):480–495
9. Cain JD, Jinapala K, Makin IW, Somaratna PG, Ariyaratna BR, Perera LR (2003) Participatory decision support for agricultural management. Sri Lanka. Agric Syst 76(2):457–482
10. Callon M (1986) The sociology of an actor-network: the case of the electric vehicle. In: Callon M, Law J, Rip A (eds) Mapping the dynamics of science and technology: sociology of science in the real world. Macmillian, London, pp 19–34
11. Callon M, Latour B (1981) Unscrewing the big Leviathan: how actors macrostructure reality and how sociologists help them to do so. In: Knorr-Cetina K, Cicourel AV (eds) Advances in social theory and methodology: toward an integration of micro- and macro-sociologies. Routledge and Kegan Paul, Boston, pp 277–303
12. Capra F (1996) The web of life: a new scientific understanding of living systems. Anchor Books, New York
13. Carley KM (2002) Computational organizational science and organizational engineering. Simul Model Pract Theory 10(5–7):253–269

14. Carroll N, Whelan E, Richardson I (2010) Applying social network analysis to discover service innovation within agile service networks. J Serv Sci 2(4):225–244
15. Carroll N (2012) Service science: an empirical study on the socio-technical dynamics of public sector service network innovation. Unpublished Ph.D. thesis, University of Limerick
16. Cartelli A (2007) Socio-technical theory and knowledge construction: towards new pedagogical paradigms? Issues Inform Sci Inform Technol 4:1–14
17. Clayton AMH, Radcliffe NJ (1996) Sustainability—a systems approach. Earthscan Publication, London
18. Coyle RG (1997) System dynamics modelling: a practical approach. Chapman and Hall, London, UK
19. Van Dam KH, Lukszo Z, Srinivasan R (2009) Agent-based decision support for failure-prone networked infrastructures. Int J Crit Infrastruct 5(4):323–339
20. Decleris M (1986) Systemic theory. Komotini, Sakkoulas, Athens
21. Dwyer C (2011) Socio-technical systems theory and environmental sustainability. In: Proceedings of SIGGreen workshop. Sprouts: working papers on information systems. Available at: http://sprouts.aisnet.org/11-3
22. Feng YY, Chen SQ, Zhang LX (2013) System dynamics modelling for urban energy consumption and CO_2 emissions: a case study of Beijing, China. Ecol Model 252:44–52
23. Fernández C, Ley E (2002) Bayesian modelling of catch in a north-west Atlantic fishery. J Roy Stat Soc: Ser C (Appl Stat) 51(3):257–280
24. Freeman L (2006) The development of social network. Empirical Press, Vancouver
25. Geels FW (2004) From sectoral systems of innovation to socio-technical systems—Insights about dynamics and change from sociology and institutional theory. Res Policy 33(6–7):897–920
26. Geels FW, Kemp R (2007) Dynamics in socio-technical systems: typology of change processes and contrasting case studies. Technol Soc 29(4):441–455
27. Gibson JJ (1979) The ecological approach to visual perception. Houghton Mifflin, Boston, MA
28. Gill J (2002) Bayesian methods: a social and behavioral sciences approach. CRC Press, Boca Raton, FL
29. De Greene KB (1988) Long wave cycles of sociotechnical change and innovation: a macropsychological perspective. J Occup Psychol 61(1):7–23
30. Hitchcock G (1993) An integrated framework for energy use and behaviour in the domestic sector. Energy Build 20:151–157
31. Hughes TP (2000) The evolution of large technological systems. In: Martin BR, Nightingale P (eds) The political economy of science, technology and innovation: Elgar reference collection. International library of critical writings in economics, vol 116
32. Iivari J, Hirschheim R (1996) Analyzing information systems development: a comparison and analysis of eight information systems development approaches. Inf Syst 21(7):551–575
33. Jarman A, Kouzmin A (1990) Decision pathways from crisis: a contingency-theory simulation heuristic for the challenger shuttle disaster (1983–1988). Contemp Crises 14 (4):399–433
34. Jayanesa HAH, Selker JS (2004) Thousand years of hydraulic civilization: some sociotechnical aspects of water management. World Water Counc, 225–262
35. Jennings NR (2000) On agent-based software engineering. Artif Intell 117(2):277–296
36. Jensen FV (2001) Bayesian networks and decision graphs. Springer, New York
37. Johnson J (2008) Science and policy in designing complex futures. Futures 40(6):521–537
38. Katz D, Kahn R (1966) The social psychology of organizations. Wiley, New York
39. Kavgic M, Mavrogianni A, Mumovic D, Summerfield A, Stevanovic Z, Djurovic-Petrovic M (2010) A review of bottom-up building stock models for energy consumption in the residential sector. Build Environ 45:1683–1697
40. Kohler N, Hassler U (2002) The building stock as a research project. Build Res Inf 30(4):226–236

41. Li YF, Li YP, Huang GH, Chen X (2010) Energy and environmental systems planning under uncertainty—an inexact fuzzy-stochastic programming approach. Appl Energy 87:3189–3211
42. Lock S (2004) The management of socio technical systems using configuration modelling. Hum Syst Manage 23(1):29–47
43. Lock S (2005) Strider: configuration modelling and analysis of complex systems. In: Proceedings of the 21st IEEE international conference on software maintenance (ICSM), 26–29 Sept 2005, pp 495–504
44. Marques FC, Dal Soglio FK, Ploeg JD (2010) Constructing sociotechnical transitions towards sustainable agriculture: lessons from ecological production of medicinal plants in Southern Brazil. In: Proceedings of innovation and sustainable development in agriculture and food, Montpellier, 28–30 June 2010
45. Masys AJ (2006) Understanding climate change through modelling and simulation: a case for verification, validation and accreditation. In: Proceedings of the 2006 IEEE EIC climate change technology conference, EICCCC 2006, Ottawa, ON
46. McIntosh BS, Jeffrey P, Lemon M, Winder N (2005) On the design of computer based models for integrated environmental science. Environ Manage 35(6):741–752
47. McNeese MD, Perusich K, Rentsch JR (2000) Advancing sociotechnical systems design via the living laboratory. In: Proceedings of the XIVth triennial congress of the international ergonomics association and 44th annual meeting of the human factors and ergonomics association, 'ergonomics for the new millennium', pp 610–613
48. Mhalas A, Kassem M, Crosbie T, Dawood N (2013) A visual energy performance assessment and decision support tool for dwellings. Visual Eng 1:7
49. Motawa IA, Banfill PF (2010) Energy-efficient practices in housing—a system dynamics approach. In: Proceedings of 18th CIB world building congress: TG62—built environment complexity, Salford, UK, May 2010, pp 44–56
50. Natarajan S, Padget J, Elliott L (2011) Modelling UK domestic energy and carbon emissions: an agent-based approach. Energy Build 43:2602–2612
51. Ogunlana S, Lim J, Saeed K (1998) Desman: a dynamic model for managing civil engineering design projects. Comput Struct 67(5):401–419
52. Olla P, Atkinson C, Gandceha R (2003) Wireless systems development methodologies: an analysis of practice using actor network theory. J Comput Inf Syst Fall 44:102–111
53. Panagiotakopoulos PD (2005) A systems and cybernetics approach to corporate sustainability in construction. Ph.D. thesis, School of the Built Environment, Heriot-Watt University, Edinburgh, UK
54. Patnayakuni R, Ruppel C (2010) A socio-technical approach to improving the systems development process. Inf Syst Front 12(2):219–234
55. Ramanna S, Skowron A, Peters JF (2007) Approximation space-based socio-technical conflict model, vol 4481. Springer, Berlin, Heidelberg
56. Ritchey T (2011) Wicked problems—social messes: decision support modelling with morphological analysis, vol 17. Springer, Berlin, Heidelberg
57. Rohracher H (2003) The role of users in the social shaping of environmental technologies. Innov Eur J Soc Sci 16(2):177–192
58. Ropohl G (1999) Philosophy of socio-technical systems. Soc Philos Technol 4(3). Available at: http://scholar.lib.vt.edu/ejournals/SPT/v4_n3html/ROPOHL.html
59. Shah AP, Pritchett AR (2005) Work-environment analysis: environment centric multi-agent simulation for design of socio-technical systems, vol 3415. Springer, Berlin, Heidelberg
60. Shipworth D (2006) Qualitative modelling of sustainable energy scenarios: an extension of the Bon qualitative input–output model. Constr Manage Econ 24(7):695–703
61. Shipworth D (2005) Synergies and conflicts on the landscape of domestic energy consumption: beyond metaphor. Paper presented at the ECEEE summer study, Mandelieu La Napoule, France
62. Sinclair MA (2007) Ergonomics issues in future systems. Ergonomics 50(12):1957–1986
63. Skyttna L (2006) General systems theory: problems, perspective, practice. World Scientific Publishing Company

64. Smajgl A, Izquierdo LR, Huigen M (2008) Modeling endogenous rule changes in an institutional context: the Adico sequence. Adv Complex Syst 11(2):199–215
65. Sterman J (1989) Testing behavioural simulation models by direct experiment. Manage Sci 33 (12):1572–1592
66. Sterman J (1992) System dynamics modelling for project management. MIT, Cambridge, MA
67. Sterman J (2000) Business dynamics: systems thinking and modelling for a complex world. Irwin McGraw-Hill, Boston
68. Sutcliffe A, Chang W, Neville RS (2007) Applying evolutionary computing to complex systems design. IEEE Trans Syst Man Cybern Part A Syst Hum 37(5):770–779
69. Thissen WA, Herder PM (2003) Critical infrastructures: challenges for systems engineering. In: Proceedings of the IEEE international conference on systems, man and cybernetics, pp 2042–2047
70. Turner B (1978) Man-made disasters. Wykeham Publications, London
71. Van Dam KH, Lukszo Z (2006) Modelling energy and transport infrastructures as a multi-agent system using a generic ontology. In: Proceedings of the 2006 IEEE international conference on systems, man, and cybernetics (SMC), The Grand Hotel, Taipei, Taiwan, pp 890–895
72. De Waal A, Ritchey T (2007) Combining morphological analysis and Bayesian networks for strategic decision support. ORiON 23(2):105–121
73. Walker GH, Stanton NA, Salmon PM, Jenkins DP (2008) A review of sociotechnical systems theory: a classic concept for new command and control paradigms. Theor Issues Ergon Sci 9 (6):479–499
74. Waterson P (2009) A critical review of the systems approach within patient safety research. Ergonomics 52(10):1185–1195
75. Williams R, Edge D (1996) The social shaping of technology. Res Policy 25:865–899
76. Wilson JR, Farrington-Darby T, Bye R, Hockey GRJ (2007) The railway as a socio-technical system: human factors at the heart of successful rail engineering. Proc Inst Mech Eng Part F 221(1):101–116
77. Wu Z, Xu J (2013) Predicting and optimisation of energy consumption using system dynamics—fuzzy multiple objective programming in world heritage areas. Energy 49:19–31
78. Yahja A, Carley KM (2005) WIZER: an automated intelligent tool for model improvement of multi-agent social-network systems. In Proceedings of the eighteenth international Florida artificial intelligence research society conference, FLAIRS 2005—recent advances in artificial intelligence, pp 44–49
79. Zadeh L (1979) A theory of approximate reasoning. Halstead Press, New York
80. Zwicky F (1969) Discovery, invention, research—through the morphological approach. The Macmillan Company, Toronto

Chapter 4
The System Dynamics Modelling Method

4.1 Introduction

In Chap. 3 of this book, the system dynamics (SD) approach was identified as one of the approaches that can be used to conceptualise the problem in the context of this book. As such, the SD modelling method and tool were applied for modelling and simulation in this book. It is important to discuss the SD approach in detail and present how it is applied in this book. Therefore, this chapter describes the theoretical background of the SD approach and its philosophical underpinnings. The chapter also gives details of the SD modelling process, including the step-by-step approach used in building the model in this book. Further, the software used to implement the SD modelling is discussed, including a description of symbols and conventions used in SD tools. Discussions of the ways in which the SD algorithms are developed conclude the chapter.

4.2 Historical Overview of the System Dynamics Approach

Historically, the SD emanates from the Massachusetts Institute of Technology's (MIT) Sloan School of Management in the 1950s when Jay W. Forrester introduced the approach in his quest to link engineering and management. This idea was conceived in order to solve complex problems considering Forrester's background in computers and feedback control systems. His ideas relating to this field of study were made known in 1956 through a seminar paper presented at the Faculty Research Seminar of the MIT's Sloan School of Management [27]. Forrester started communicating his ideas by offering fierce criticism of economic models and their assumptions. Forrester's main criticism of economic models is hinged on the fact that those economic models [20]:

M. G. Oladokun and C. O. Aigbavboa, *Simulation-Based Analysis of Energy and Carbon Emissions in the Housing Sector*, Green Energy and Technology,
https://doi.org/10.1007/978-3-319-75346-1_4

- fail to adequately reflect the loop structure that make up economic systems and as such neglect leads to exclude inherent properties of closed loops such as resistance to change, accumulations, and delays;
- are incapable of including flows of goods, money, information, and labour in one single interrelated model;
- exclude changing mental attitudes that affect and explain economic processes;
- use linear equations for describing systems;
- offer a restriction in building models as constrained by the capacity for manipulating numerical data and solving the equations;
- rely on multiple regression analysis for obtaining coefficients for equations that define economic behaviour; and
- lack reflection on the very assumptions underlying the models.

After the arguments, Forrester then went ahead to present his approach that was hinged on servomechanisms and differential equations, the techniques that were argued to be largely underutilised then [22]. Premise on the above, Forrester considered a new approach to understanding the firm and economy, as reported in Forrester [22] by proposing a new kind of model that incorporates (1) dynamic structure; (2) information flows; (3) decision criteria; (4) nonlinear systems; (5) differential equations; (6) incremental changes in variables; (7) model complexity; (8) symbolism and correspondence with real counterparts; and (9) structure over coefficient accuracy.

Forrester then made more advances on the technique, which led to the publication of an article entitled 'Industrial Dynamics: A Major Breakthrough for Decision Makers in the Harvard Business Review' [23] in 1958. In this article, Forrester was able to shape his ideas and since then, the SD approach has emerged as a multidisciplinary field of study that deals with the analysis of complex systems. The approach has indeed remained a powerful and well-established methodology and tool for modelling and understanding feedback structure in complex systems [2, 49, 66]. The approach, according to Coyle [8], '*...deals with the time-dependent behaviour of managed systems with the aim of describing the system and understanding, through qualitative and quantitative models, how information feedback governs its behaviour, and designing robust information feedback structures and control policies through simulation and optimisation*', and was grounded in theory of modern feedback control and nonlinear dynamics. Furthermore, it is built on 'cause-and-effect' relations among different variables influencing the system under investigation [49] and is indeed a '*...method to enhance learning in complex systems*' [59]. The next section gives the theoretical and philosophical underpinnings of the SD approach.

4.3 Theoretical and Philosophical Underpinnings of the System Dynamics Method

In order to address the philosophical issues regarding the SD, it would be necessary to first ask: What is system dynamics? Is it a paradigm, philosophy, theory, methodology, method, or a set of techniques or tools? Answering these questions has indeed generated a high-level debate in the SD research circle as different system dynamicists and researchers have attributed different names to it. For example, some system dynamicists see SD as a theory [18, 28], a (group of) method (s) [7, 32, 36, 59, 62], a methodology [56, 60], a field of study [9, 50], a tool [2, 33, 66], a paradigm [1, 25, 34, 36, 37, 48, 55], and a host of other nouns/ nomenclatures? It is therefore necessary to place the SD approach correctly within the philosophical research paradigm space by assessing its premises regarding how it is grounded in philosophical foundations. Is it a positivist, interpretivist, or pragmatist philosophical research paradigm?

In answering the above question, Jackson [31], for example, considers SD models to be a representation of an assumed "objective" real world, thereby labelling the SD approach as a "hard" approach as is further reflected in the work of Flood and Jackson [18]. This means that the SD approach is placed within the one-dimensional array of *positivism*. This categorisation cannot be utterly refuted as Zagonel [65] gives some examples of SD practices that illustrate this position, namely policy engineering, optimisation-based simulation, purely quantitative SD, and micro-world modelling. Following this position, Pruyt [47] contends that the positivist SD assumes an ontological position that models systems corresponding to the existing systems in the real world. Also, the epistemological position of this paradigm assumes that the stock and flow diagrams and causal loop diagrams are good objective representations of the external reality. This position presumes that the resulting quantitative SD simulation presents an approach to replicate the dynamics of these real-world systems.

Yet other studies have criticised viewing the SD approach as a purely objective (interpretivist) research paradigm. Forrester[1] himself [20] stressed that '…*a model can be useful if it represents only what we believe to be the nature of the system under study…we are forced to commit ourselves on what we believe is the relative importance of various factors. We shall discover inconsistencies in our basic assumptions…Through any of these we learn*'. The justification to disprove the arguments of critics who mistakenly placed the SD approach within the purely positivism space is apparent from this statement, namely that SD is more than simulations alone. This is because it qualitatively captures the mental models of practitioners involved in SD modelling. Further to this, a number of system dynamicists [12, 24, 59] have argued that the central concept driving the SD approach lies in 'mental models', which presumably are purely subjective.

[1]Jay W. Forrester is the pioneer of the SD approach.

However, the ability and adequacy of 'mental models' that are labelled as the core of the SD approach to depict reality accurately have generated a kind of debate within the research circles [12]. Apparently, it would be necessary to further explore the idea of mental models in order to see on how they have helped shape the SD approach. Sterman [59], for example, expressly refers to 'mental models' as '*...our beliefs about the networks of causes and effects that describe how a system operates, along with the boundary of the model (which variables are included and which are excluded) and the time horizon we consider relevant...Most of us do not appreciate the ubiquity and invisibility of mental models, instead believing naively that our senses reveal the world as it is. On the contrary, our world is actively constructed (modelled) by our senses and brain*'. Further, Lane [34] considers 'mental models' as desired systems conceived and existing in the mind of the modeller. Pruyt [47] describes both the ontological and epistemological positions of interpretivist SD. He sees the ontological position as relativist and the epistemological position as subjective.

From the foregoing, it is yet unclear whether it is adequate to place the SD approach in any of the research paradigm divides above. This then necessitates the need to assess the pragmatist SD. Pruyt [47], for example, argues that most SD practices contain pragmatist elements within them as they tend to reflect the characteristics of both the positivist and interpretivist paradigms. That is, the ontological or epistemological assumption of the SD approach is more realist and objective—as is the case during the simulation phase—and sometimes more nominalist and subjective—as is the case during the modelling and interpretation phases [47]. Barton [5] suggests that the philosophical underpinning of the systems thinking stage of SD approach (referred to as the modelling phase by [47] lies in pragmatism. Stemming from the preceding arguments, this book therefore places the SD approach within the pragmatist research paradigm frame. This is based on considering the fact that the SD approach combines both the system thinking (interpretivist paradigm) and simulation (positivist paradigm) aspects to develop the methodology.

4.4 System Dynamics as a Multidisciplinary Modelling Approach

In order to put SD in context as a multidisciplinary modelling approach, it is important to further discuss its application. Undoubtedly, SD is seen as a powerful tool which has the capability of being used in many fields of study, especially where the behaviour of a system and its understanding are to be studied [49]. For example, SD has found application in economics and finance [21, 27], resource management [14, 53], education [31], health [30, 38], production management [16, 54], project management [44], public policy and management [10, 58], strategy [4, 29], energy and environment [3, 64], and within the energy consumption and carbon emissions domain [17, 41, 42, 63]. Additionally, the SD approach has not only been used for energy efficiency; it has been developed and used in a number of contexts as demonstrated by the work of Davis and Durbach [11], Motawa and Banfill [40], and Dyner et al. [13], among others. According to this, the benefit lies with SD application to energy policy based on the studies of Oladokun and Adewuyi [46], Chi et al. [6], Naill [43], and Ford [19], among others. All these studies reinforce the application of the SD approach to energy policy evaluation and as such the SD perspective is being used for the simulation of energy policy within the housing sector in this book.

4.5 Procedures for Using the System Dynamics Method

Within the body of SD literature, there are a number of stages that have been identified by different SD modellers as procedure for implementing the SD modelling and simulation. Some of these stages often overlap. For example, the work of Randers [48] suggested four stages as procedure for implementing the SD modelling and simulation. These stages include model conceptualisation, formulation, testing, and implementation. Contrary to this suggestion, Wolstenholme [62] collapsed these stages into two, namely diagram conceptualisation, and analysis and simulation. Rather than collapsing the stages into two, Sterman [59] extended these stages as suggested by both Randers [48] and Wolstenholme [62] to include five distinct stages, namely problem articulation, dynamic hypothesis, model formulation and simulation, testing, and policy formulation and evaluation. The work of Robert et al. [57], Richardson and Pugh [55], and Ranganath and Rodrigues [49] suggests that for any SD modelling and simulation, the process should incorporate stages that include problem identification, system conceptualisation, model formulation, analysis of model behaviour, model evaluation, policy analysis and improvement, and policy implementation. Martinez-Moyano and Richardson [35] extend what Robert et al. [57], Richardson and Pugh [55], and Ranganath and Rodrigues [49] consider being the main stages of SD modelling efforts.

Owing to the nature of the research problem in this book, the SD modelling procedure as applied in Chap. 5 of this book consists of problem identification and definition, system conceptualisation, model formulation, model behaviour analysis, model testing and validation, and policy formulation and analysis as shown in Table 4.1. These are then mapped into four main stages as depicted in Fig. 4.1 based on the work of Oladokun and Odesola [45]. The modelling process for the research includes the timeline, major tasks performed, and the methodology employed to achieve each of the tasks performed.

Table 4.1 Approaches to the system dynamics modelling process

Randers [48]	Roberts et al. [57]	Wolstenholme [62]	Richardson and Pugh [55]	Sterman [59]	Ranganath and Rodrigues [49]	Martinez-Moyano and Richardson [35]	This book
Conceptualisation	Problem definition	Diagram construction and analysis	Problem identification and definition	Problem articulation	Problem identification and definition	Problem identification and definition	Problem identification and definition
	System conceptualisation		System conceptualisation	Dynamic hypothesis	System conceptualisation	System conceptualisation	System conceptualisation
Formulation	Model representation	Simulation phase (stage 1)	Model formulation	Formulation	Model formulation	Model formulation	Model formulation
	Model behaviour		Analysis of model behaviour				Model behaviour analysis
Testing	Model evaluation	Simulation phase (stage 2)	Model evaluation	Testing	Simulation and validation	Model testing and validation	Model testing and validation
			Policy analysis		Policy analysis and improvement	Model use, implementation, and dissemination	
Implementation	Policy analysis and model use	Policy analysis and evaluation	Model use or implementation	Policy formulation and evaluation	Policy implementation	Design of learning strategy/ infrastructure	Policy formulation and analysis

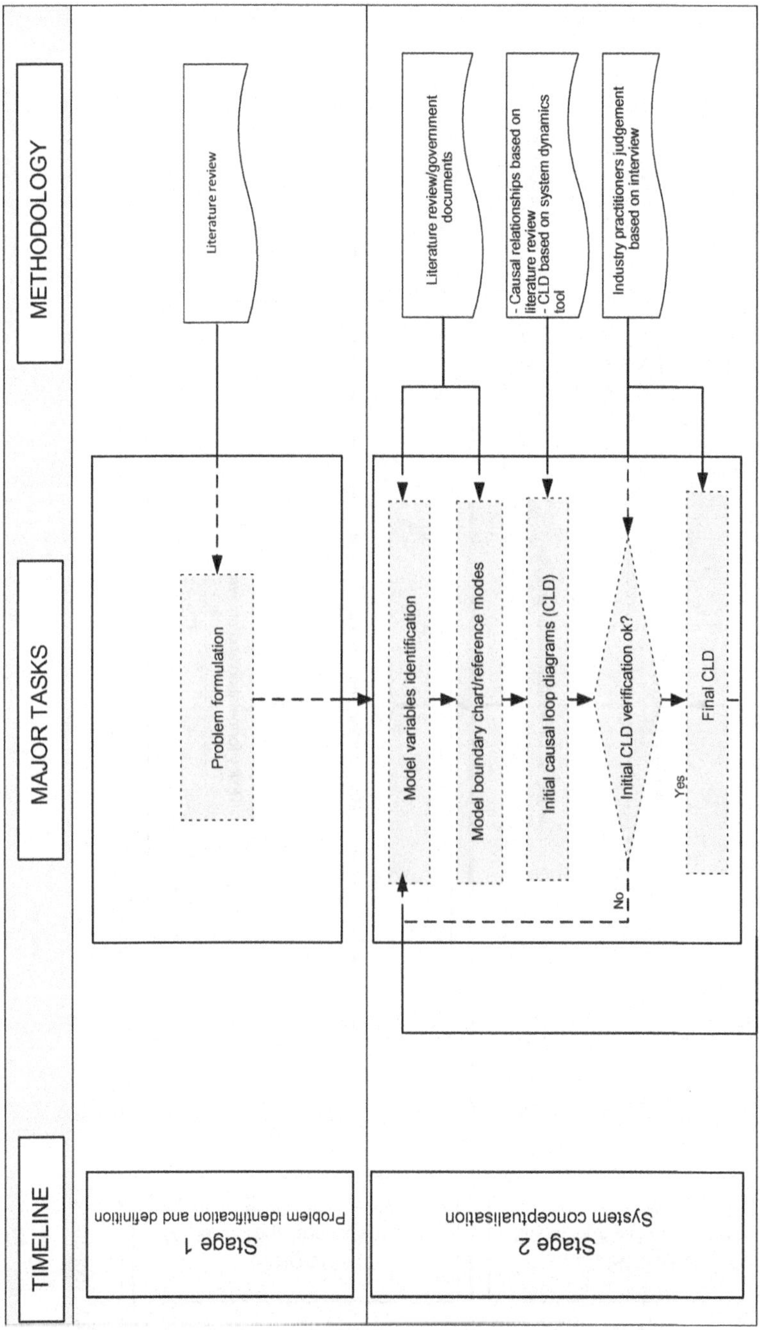

Fig. 4.1 Modelling process [45]

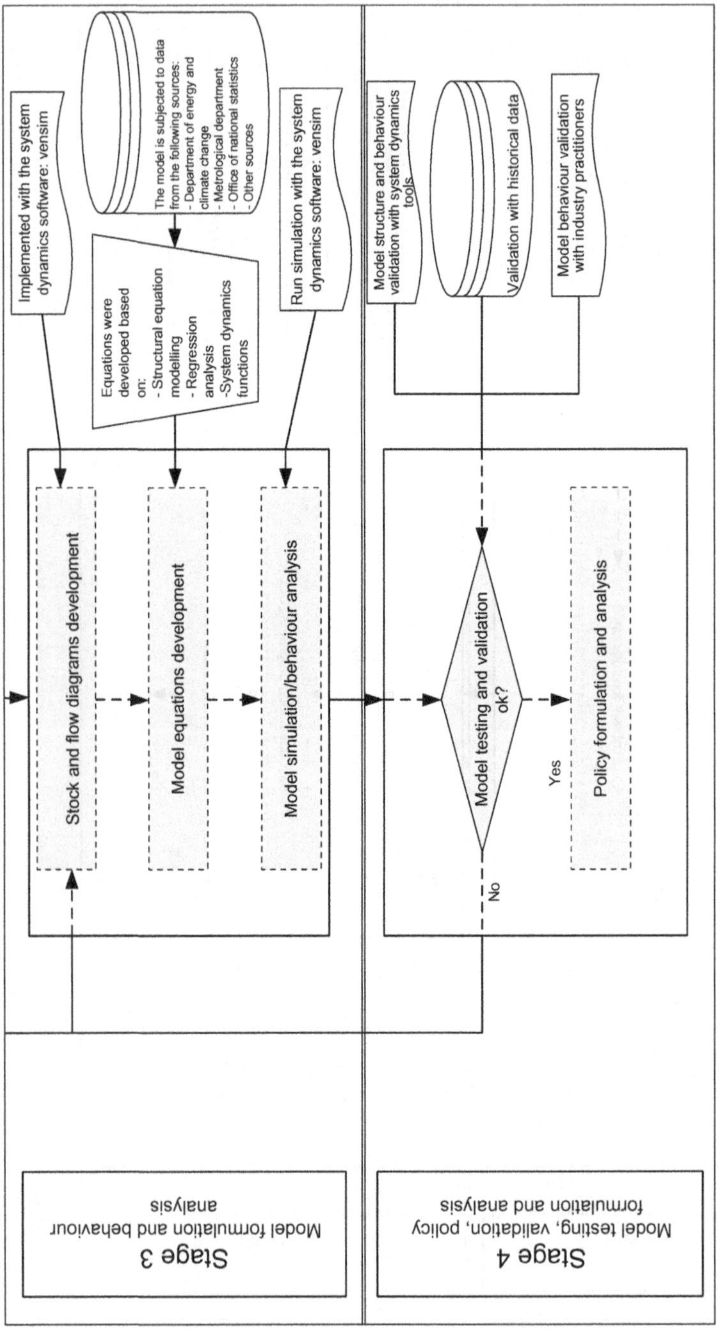

Fig. 4.1 (continued)

4.5.1 Stage 1: Problem Identification and Definition

As shown in Fig. 4.1, the first stage identifies the research problem and places it in context by reviewing relevant studies as identified in the literature. This is necessary in order to contextualise the problem the SD approach intends to solve. The rationale for the study in this book has been established in Chap. 1 of this book.

4.5.2 Stage 2: System Conceptualisation

As displayed in Fig. 4.1, the second stage as firmed up for the SD approach in this book has to do with the system conceptualisation. This stage does not necessarily require the modeller to catalogue quantitative data in order to conceptualise the problem. However, the modeller mainly focuses on reviewing extant literature on the problem and how the mental knowledge of the experts in the field of study can be captured in a bid to develop the initial characterisation of the problem. Sterman [59] reinforces the importance of this exercise when he submits that modellers usually have discussions with relevant stakeholders within the frame of the research, which are supplemented by reviewing extant literature, interviews, and direct observations or participation.

The exercise above is performed in such a way so as to identify what the model variables are. This is done by reviewing extant literature, government documents, and direct observation. Within the frame of the research being reported in this book, extant literature was reviewed and a number of government reports and documents from different sources were reviewed as well. For example, there were many documents and reports from the Department of Energy and Climate Change that were retrieved and analysed. This analysis resulted in the model boundary and reference modes for key variables in order to know the variables to be included in the model as well as the trend of the key variables over time. Furthermore, the initial causal loop diagrams (CLDs) of the variables identified were established to give the mental model in terms of a feedback structure for the system under study. These CLDs can be drawn with the use of appropriate SD software as indicated in Sect. 4.6 of this book.

It is important to clarify that the initial CLDs formulated merely represent sets of dynamic hypotheses. These initial CLDs are based on the knowledge elicitation of the modeller (authors) as informed by extant literature in the field. Input from the experts on the subject is then captured in the form of knowledge elicitation by having discussions with a number of them (the number of experts to be included needs to be justified and grounded in literature). The purpose of that exercise is to verify and validate the initial CLDs that were based on the idea of the system dynamicist alone which in turn was based on the literature review and content analysis. In this exercise, there may be removal, addition of some causal links, and even addition of extra variables until the final CLDs are achieved. It is noteworthy that the final CLDs at this stage are the system thinking aspect of the SD approach and do not necessarily indicate the SD modelling. The SD modelling will transform the CLDs into what are called the stock and flow diagrams (SFDs) and then relate the variables together using equations and different algorithms. Details about the CLDs are explored in Chap. 5 of this book.

4.5.3 Stage 3: Model Formulation and Behaviour Analysis

In this stage, a detailed model structure is given together with model parameter values. Depending on the research problem, this stage mostly/largely used to contain elements of quantitative data as the case is in this book. In some cases, however, formulation of qualitative concepts is done here as well, as with the case in this book. In fact, one of the main strengths of the SD approach is its ability to combine both the qualitative and quantitative research approaches as debated in Sect. 4.3 of this chapter. Richardson and Pugh [52] add to the qualitative model formulation by suggesting that '*…the modeller may wish to represent a concept explicitly. To do this requires the invention of units and a measurement scale, and consistent treatment throughout the model*'.

The use of qualitative data sources for model formulation has received criticism within the research circles as a number of researchers have raised some concerns about the applicability of qualitative data in equations formulation. The defence against this query has been that the tenet of the SD approach is founded on taking advantage of the mental models of experienced practitioners within the subject area. For example, Richardson [51] addresses the issue of qualitative mapping and formal modelling. In the same vein, Sterman [59] contends that '*…omitting structures or variables known to be important because numerical data are unavailable is actually less scientific and less accurate than using your best judgement to estimate their values*'. Taking all these into cognisance, the model in this research book is formulated with the use of both qualitative and quantitative data sources as previously mentioned.

Figure 4.1 shows the third stage of the modelling process, which has to do with model formulation and behaviour analysis. Formulating the model has to do with transforming the CLDs into SFDs. The SFDs reveal the behaviour of the model pictorially, indicating the accumulation (stock) and flow (rate) as achieved and implemented within the SD platform. It is noteworthy that mere CLDs or SFDs do not automatically lead to a SD simulation. The SFDs can only be modelled and simulated once all the variables in the model have been linked together via appropriate algorithms in the form of equations. As such, model algorithms are developed using a number of different approaches involving both qualitative and numerical data. For example, algorithms can be developed using the SD functions and molecules within the SD software to capture both qualitative and quantitative data. Additionally, approaches such as the regression analysis and structural equation modelling, among others, can be explored. The model in this book as shown in the succeeding chapters is subjected to various data sources in the UK such as the Department of Energy and Climate Change (DECC), the Met Office, and the ONS in order to develop the algorithms. Also, data collected via interviews during the knowledge elicitation and from the literature are qualitatively inputted into the model through 'look-ups' as can be found in any SD software. Once the system configuration is found to be satisfactory, the simulation is then run using a time horizon from 1970 to 2050 with a year time step and the use of the Euler form of integration type. Model behaviour is then analysed and discussed accordingly. Full details about model formulation are presented in Chap. 5, and model behaviour analysis is addressed in Chap. 6.

4.5.4 Stage 4: Model Testing and Validation, and Policy Formulation and Analysis

Fourth stage, as shown in Fig. 4.1, concludes the modelling process in building an SD model of the problem addressed in this book. This stage consists of model testing and validation, and policy formulation and analysis. The studies of Forrester and Senge [26] and Sterman [59] provide a comprehensive array of different tests to be performed in order to have SD models validated. The studies broadly divide the tests into two to include (1) structure verification and (2) behaviour verification (see Sect. 6.5 for a list all the tests based on Sterman [59]). Forrester and Senge [26] highlight the importance of this exercise when they commented on structure verification, for example, that '*…the model must not contract knowledge about the structure of the real system…In most instances, the structure verification test is first conducted on the basis of the model builder's personal knowledge and is then extended to include criticisms by others with direct experience from the real system*'. In the same vein, Randers [48] establishes how the test should be performed and who should take part in it by suggesting that '*…the modeller should not restrict himself to the small fraction of knowledge available in numerical form fit for statistical analysis. Most human knowledge takes a descriptive non-quantitative form…Model testing should draw upon all sources of available knowledge*'. These statements emphasise the importance of model validation with experts and practitioners in the field of study.

In this book, model validation is seen as a way of testing and verifying the model structure, behaviour, and sensitivity analysis using the SD functions within the software. Furthermore, validation against historical data is done where applicable according to any available historical data. With this, the predictive strength of the model and its capability to project into the future are evaluated statistically. Additionally, interviews can be conducted to ensure that the results in terms of model behaviour are reviewed by the experts and industry practitioners. This may be done with the experts and industry practitioners who had previously been contacted at second stage as firmed up by this modelling approach. The main purpose of this exercise is to establish whether the trends witnessed in terms of the behaviour of the key variables reflect their expectations since they have considerable experience in the subject. This exercise will increase the credibility of the developed model.

After the model testing and validation have been completed satisfactorily, policy formulation can then be done in order to conduct some policy analysis based on the policy levers that have already been introduced within the body of the model. Here, simulation of different policy scenarios is performed (Chap. 7).

4.6 System Dynamics Modelling Tools

Modelling and simulation of the SD models are implemented using a number of platforms or software. Of these platforms or software, the following are discussed in this book: DYNAMO, Powersim, STELLA/iThink, AnyLogic, and Vensim, among others. Based on the extension of Eberlein's [15] work, brief details about some of the software are presented as follows:

- *DYNAMO*: Within the SD modelling, Dynamic Models (DYNAMO) represents the first simulation language developed in the field. The software was developed at Massachusetts Institute of Technology (MIT) by Jack Pugh in the 1960s. It can run on personal computers under the DOS/Windows platform. It provides an equation-based modelling environment.
- *Powersim*: Powersim became popular in the mid-1980s as a result of the research sponsored by the Norwegian government in a bid to support and improve high school students in the use of SD models. The output of the research gave birth to the object-oriented simulation-based games that are primarily used for education. Powersim can be used in a Windows-based environment for creating SD models with the ability of facilitating interactive games or learning experiments.
- *STELLA* (Structural Thinking Experimental Learning Laboratory with Animation): STELLA was first inaugurated on the Macintosh in the 1980s, but is now available on Windows. The software provides a graphical interface for the development of SD models. It supports a series of tools in the model development which allows easy access to equation writing.
- *AnyLogic*: AnyLogic software was developed in the 2000s as organised by the Distributed Network Computer research group at St. Petersburg Technical University. The software has the capability of supporting the SD, discrete event simulation, and ABM. The software can work on Windows, Macintosh, and Linux as well.
- *Vensim*: Vensim was developed by Ventana Systems around the mid-1980s primarily for use in consulting projects. The software became commercially available in 1992 and works well in both Windows and Macintosh. Vensim fully supports SD modelling with flexible graphical representations without any form

of clustering on the interface. The software contains a panoply of tools for model analysis and testing, and the results can be visualised instantly on invoking SyntheSim. Vensim has the capability of using data and calibrating the same. The software also has the capability of being linked to other software such as C, C ++, and Visual Basic, among others. Also, other SD modelling software such as Powersim and STELLA can be easily converted to Vensim.

According to Coyle [8], before settling for any of the software, it is necessary to assess the software package based on its basis in fundamental SD theory; the ease with which it can be used; the support it gives to model building; the extent to which models can be documented and explained to a sponsor; the facilities it has for debugging a model; the ease of making/doing/conducting experiments and producing output; and the scope of its facilities for policy design. When all the identified SD software is assessed against the above set of criteria together with their capabilities as explained above, they all have what it takes to be used. However, Vensim is chosen as the SD software for the modelling in this book because of its flexible graphical representations, which aid its clarity in presenting the CLDs. Also, its ability to incorporate optimisation, the method that was derived from numerical mathematics, presents an added advantage over all other software, motivating its use.

Within the Vensim software, there is an array of symbols used to create SD model sketches. In addition, Vensim utilises a workbench that allows the modeller to build and analyse a model and its accompanied datasets [61]. According to Ventana [61], the workbench comprises a menu, a model, a toolbar, a variable, control panel, one or more toolsets, output windows, and model building windows. An example of this workbench is shown in Fig. 4.2, displaying the different tools and symbols used in Vensim.

Within the Title Bar, the Model (Model_Test-06122013.mdl) and Variable (average household size) are named and displayed. Immediately after the Title Bar is the Menu, which is followed by the Main Toolbar for the software. Below the Main Toolbar is the Sketch Toolset used in sketching the model. To the left of the screen is the Analysis Toolset for performing the analysis of the Model Output as shown in another window in Fig. 4.2. Also, the Control Panel is shown in another window displaying the loaded and available datasets. The Build Window reflects the workbench where sketches related to the model are created and edited.

Additionally, Table 4.2 shows some of the conventions used in Vensim especially in depicting different types of variables within the SD models such as stock, flow, and the like. Some of the other conventions are highlighted as well. The description of these terminologies is given in Table 4.2.

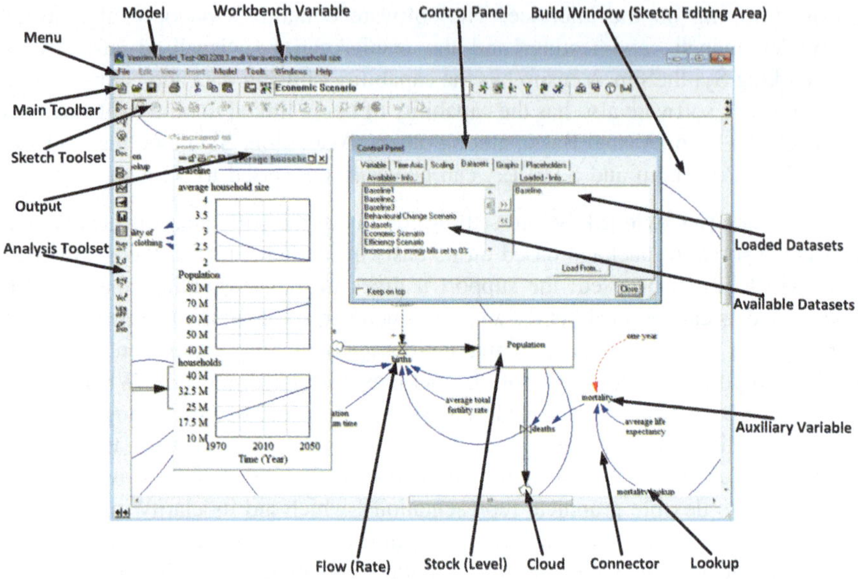

Fig. 4.2 Vensim workbench (Source: [48])

Table 4.2 Conventions used in SD modelling

Name	Description
Stock (level)	A stock can be defined as a structural term for anything that accumulates
Flow (rate)	If stocks/levels are bathtubs, then flows/rates are pipes that feed and drain them
Cloud	A cloud is an infinite reservoir representing the boundary. The capacity of cloud is so great that it makes no sense to worry about filling or draining it
Connector (arrow)	A connector/arrow is used to link the variables in the model together
Auxiliary variables	These are computed from Levels, Constants, Data, and other Auxiliaries. Auxiliary variables have no memory, and their current values are independent of the values of variables at previous times
Look-ups	They are used in specifying arbitrary nonlinear relationships in Vensim

Adapted with modifications from [39]

4.7 System Dynamics Algorithms Development

Variables (especially levels, rates, auxiliary, among others) are to be related to one another in the form of equations. Developing the main algorithms in contemporary SD paradigm involves using an array of functions embedded within the SD modelling software. For example, within the Vensim software, algorithms are formed with the use of simple functions such as addition, subtraction, multiplication, and

Table 4.3 Sample relationship developed from regression analysis (Source: [48])

Model		Unstandardised coefficients		Standardised coefficients	*t*	Sig.
		B	Std. error	Beta		
1	(Constant)	−3.436E8	4,538,500.746		−75.712	.000
	Population	.067	.017	.056	3.886	.000
	Time	182,057.612	2746.898	.947	66.278	.000

Dependent variable: Households
(Source: [48])

division. Also, special functions are used within the SD modelling platform software (e.g. Vensim) to implement some computational tasks in the model. For example, there are functions such as DELAY, FORECAST, IF THEN ELSE, RAMP, SMOOTH, and STEP, among others. The ability to utilise these functions is still subject to the availability of the requisite data or correct parameter assumptions to implement these tasks and then successfully drive the model. Where there is a lack of empirical data, qualitative data are gleaned by interviews, for example, through knowledge elicitation from the experts [59]. In this book, SD functions within the Vensim software are used to mimic the reference modes[2] that are illustrating the problem. On the other hand, where there is evidence of empirical data, the modeller can take the advantages of some other methods of establishing the relationships among the variables in the model (e.g. regression analysis).

In the context of this book, both qualitative and quantitative data were used to develop the model algorithms as given in the preceding sections. As a result of this, a combination of regression analysis, structural equation modelling, and predefined equations (e.g. SAP algorithms) in some aspects are used in addition to special functions within the Vensim software. Table 4.3 gives an example of a regression-based equation developed for the relationship between population and households. Also, an example of an algorithm based on structural equation modelling with a combination of Vensim special function developed is shown in Eq. 5.25 of Chap. 5. Furthermore, examples of equations based on SAP algorithms are shown in Eqs. 5.7–5.14 of Sect. 5.6.2 of Chap. 5.

4.8 Summary

The SD approach was selected as an appropriate modelling platform for the research reported in this book. The chapter then discussed the SD approach in full, starting with the historical overview of the approach. The chapter underpinned the SD approach both in theory and philosophy. As such, the SD approach was

[2]Reference modes are elaborately discussed in Sect. 5.4 of Chap. 5.

identified within the frame of the pragmatic paradigm. This indicates that any study involving SD falls within the purview of qualitative and quantitative research strategies. The chapter also succeeds in showing the SD approach as a multidisciplinary modelling method that has been used in many fields. Additionally, the chapter discussed the four stages involved in the SD modelling process, namely problem identification and definition; system conceptualisation; model formulation and behaviour analysis; model testing and validation, and policy formulation and analysis stages. Furthermore, DYNAMO, Powersim, STELLA/iThink, AnyLogic, and Vensim, some of the software under which SD can be implemented, were briefly discussed. The chapter indicated that Vensim was used as the modelling software for the book. This is because of its flexible graphical representations which aid its clarity in presenting the CLDs as well as its ability to incorporate optimisation. The discussion of the methods used in the development of the relationships (algorithms) among the model variables concludes the chapter. These methods include the use of SD functions within the Vensim software, regression analysis, and SEM as well as other established equations such as those provided in SAP algorithms. The next chapter discusses the application of the SD approach for modelling and simulating the energy and carbon emissions problem in housing using the UK housing sector as example.

References

1. Andersen D (1980) How differences in analytic paradigms can lead to differences in policy conclusions. In: Elements of the system dynamics method. MIT Press/Wright-Allen series in system dynamics. MIT Press, Cambridge, pp 61–75
2. Ansari N, Seifi A (2013) A system dynamics model for analysing energy consumption and CO_2 emission in Iranian cement industry under various production and export scenarios. Energy Policy 58:75–89
3. Balnac K, Bokhoree C, Deenapanray P, Bassi AM (2009) A system dynamics model of the Mauritian power sector. In: Proceedings of the 27th international conference of the system dynamics society, 26–30 July, Albuquerque, NM, USA
4. Barabba V, Huber C, Cooke F, Pudar N, Smith J, Paich M (2002) A multimethod approach for creating new business models: the general motors onstar project. Interfaces 32(1):20–34
5. Barton J (2002) Pragmatism, systems thinking and system dynamics. In: Proceedings of the 20th international conference of the system dynamics society, July 28–Aug 1, Palermo, Italy
6. Chi KC, Nuttall WJ, Reiner DM (2009) Dynamics of the UK natural gas industry: system dynamics modelling and long-term energy policy analysis. Technol Forecast Soc Chang 76:339–357
7. Coyle RG (1977) Management system dynamics. Wiley, Chichester
8. Coyle RG (1997) System dynamics modelling: a practical approach. Chapman & Hall, UK
9. Coyle RG (2000) Qualitative and quantitative modelling in system dynamics: some research questions. Syst Dyn Rev 16:225–244
10. Dangerfield BC (2006) A system dynamics model for economic planning in sarawak. In: Proceedings of the international system dynamics conference, Nijmegen, The Netherlands
11. Davis S, Durbach I (2010). Modelling household responses to energy effciency interventions via system dynamics and survey data. ORiON: J ORSSA, 26(2):79–96

12. Doyle JK, Ford DN (1998) Mental models concepts for system dynamics research. Syst Dyn Rev 14:3–29
13. Dyner I, Smith R, Pena G (1995) System dynamics modelling for energy efficiency analysis and management. J Oper Res 46:1163–1173
14. Dyson B, Chang NB (2005) Forecasting municipal solid waste generation in a fast-growing urban region with system dynamics modelling. Waste Manage 25:669–679
15. Eberlein B (2013) System dynamics software info. Available from http://www.vensim.com/sdmail/sdsoft.html. Accessed 14 Mar 2013
16. Ellis RE (2001) The use of system dynamics in a large-scale system engineering effort: case study of the deepwater project. In: Proceedings of the international system dynamics conference, Atlanta, GA, USA
17. Feng YY, Chen SQ, Zhang LX (2013) System dynamics modelling for urban energy consumption and CO_2 emissions: a case study of Beijing, China. Ecol Model 252:44–52
18. Flood R, Jackson M (1991) Creative problem solving. Wiley, Chichester
19. Ford A (1983) Using simulation for policy evaluation in the electricity utility industry. Simulation 40(3):85–92
20. Forrester JW (1961) *Industrial dynamics*. MIT Press, Cambridge, Massachusetts
21. Forrester JW (1971) World dynamics. Wright-Allen Press, Cambridge
22. Forrester JW (1975) Industrial dynamics—after the first decade. Collected papers of Jay W Forrester. Wright-Allen Press, Cambridge, pp 133–150
23. Forrester JW (1975) Industrial dynamics: a major breakthrough for decision makers. Collected papers of Jay W Forrester. Wright-Allen Press, Cambridge, pp 1–29
24. Forrester JW (1975) Counter intuitive behaviour of social systems. Collected papers of Jay W Forrester. Wright-Allen Press, Cambridge, pp 211–244
25. Forrester J (1994) System dynamics, systems thinking, and soft OR. Syst Dyn Rev 10(2–3):245–256
26. Forrester JW, Senge PM (1980) Tests for building confidence in system dynamics models. In: Legasto AA, Forrester JW, Lyneis JM (eds) system dynamics. North-Holland, Amsterdam
27. Forrester, JW (2003) Dynamic models of economic systems and industrial organisations. *Sys Dyn Rev*, 19(4), 331–345
28. Ghaffarzadegan N, Tajrishi AT (2010) Economic transition management in a commodity market: the case of the Iranian cement industry. Syst Dyn Rev 26(2):139–161
29. Hitchcock CR, Salmon WC (2000) Statistical explanation. In: Newton-Smith WH (ed) A companion to the philosophy of science. Blackwell Publishers, Malden
30. Homer J (1996) Why we iterate: Scientific modelling in theory and practice. Syst Dyn Rev 12 (1):1–19
31. Homer, J (1997) Structure, data, and compelling conclusions: Notes from the field. *Sys Dyn Rev*, 13(4), 293–309
32. Homer J, Milstein B, Dietz W, Buchner D, Majestic E (2006) Obesity population dynamics: exploring historical growth and plausible futures in the U.S. In: Proceedings of the international system dynamics conference, Nijmegen, The Netherlands
33. Jackson M (2003) Systems thinking: creative holism for managers. Wiley, Chichester
34. Lane, D. (1999). Social theory and system dynamics practice. *Euro J of Oper Res*, 113(3), 501–527
35. Lane D (2001) Rerum cognoscere causas: part I—how do the ideas of system dynamics relate to traditional social theories and the voluntarism/determinism debate? Syst Dyn Rev 17 (2):97–118
36. Luna-Reyes LF, Andersen DL (2003) Collecting and analysing qualitative data for system dynamics: methods and models. Syst Dyn Rev 19:271–296
37. Maani K, Maharaj V (2004) Links between systems thinking and complex decision making. Syst Dyn Rev 20(1):21–48
38. Martinez-Moyano IJ, Richardson GP (2013) Best practices in system dynamics modelling. Syst Dyn Rev 29(2):102–123

39. Meadows D (1980) Elements of the system dynamics method. Wright-Allen series in system dynamics. MIT Press, Cambridge
40. Meadows D, Robinson J (1985) The electronic oracle. Computer models and social decisions. Wiley, Chichester
41. Milstein B, Homer J, Hirsch G (2010) Analysing national health reform strategies with a dynamic simulation model. Am J Public Health 100(5):811–819
42. Morecroft JDW (1988) System dynamics and microworlds for policymakers. Eur J Oper Res 35:301–320
43. Motawa IA, Banfill PF (2010) Energy-efficient practices in housing—a system dynamics approach. In: Proceedings of 18th CIB world building congress: TG62—built environment complexity, May, Salford, UK, pp 44–56
44. Motawa I, Oladokun M (2015) A model for the complexity of household energy consumption. Energy Build 87:313–323
45. Motawa I, Oladokun M (2015) Dynamic modelling for sustainable dwellings. Eng Sustain 168(4):182–190. https://doi.org/10.1680/ensu.14.00051
46. Naill RF (1992) A system dynamics model for national energy policy planning. Syst Dyn Rev 8:1–19
47. Ogunlana S, Lim J, Saeed K (1998) Desman: a dynamic model for managing civil engineering design projects. Comput Struct 67(5):401–419
48. Oladokun MG (2014) Dynamic modelling of the socio-technical systems of household energy consumption and carbon emissions. PhDthesis, Heriot-Watt University, Edinburgh, UK
49. Oladokun MG, Odesola IA (2015) Household energy consumption and carbon emissions for sustainable cities—a critical review of modelling approaches. Int J Sustain Built Environ 4:231–247. https://doi.org/10.1016/j.ijsbe.2015.07.005
50. Oladokun MG, Adewuyi TO (2016) A new integrated and holistic modelling approach for energy performance analysis of residential buildings. J Environ Des Manag 8(1&2):132–144
51. Pruyt E (2006) What is system dynamics? A paradigmatic inquiry. In: Proceedings of the 2006 international conference of the system dynamics society, July 23–27, Nijmegen, The Netherlands
52. Randers J (1980) Guidelines for model conceptualization. In: Elements of the system dynamics method. MIT Press/Wright-Allen series in system dynamics. MIT Press, Cambridge, pp 117–139
53. Ranganath BJ, Rodrigues LLR (2008) System dynamics: theory and case studies. International Publishing House, New Delhi
54. Richardson GP (1991) Feedback thought in social science and systems theory. Pegasus Communications, Waltham
55. Richardson GP (1996) Problems for the future of system dynamics. Syst Dyn Rev 12(3):141–157
56. Richardson GP, Pugh ALIII (1981) Introduction to system dynamics modeling with DYNAMO. Productivity Press, Cambridge
57. Rehan R, Knight MA, Haas CT, Unger AJA (2011) Application of system dynamics for developing financially self-sustaining management policies for water and wastewater systems. Water Res 45:4737–4750
58. Repenning NP, Sterman JD (2001) Nobody ever gets credit for fixing problems that never happened: creating and sustaining process improvement. Calif Manag Rev 43(4):64–88
59. Richardson GP, Pugh A (1999) Introduction to system dynamics: modelling with DYNAMO. MIT Press, Cambridge
60. Roberts EB (1978) System dynamics—an introduction. In: Roberts EB (ed) Managerial applications of system dynamics. Pegasus Communications Inc., Waltham
61. Roberts N, Anderson D, Deal R, Grant M, Shaffer W (1983) Introduction to computer simulation: a system dynamics modelling approach. Addison-Wesley, Reading
62. Rouwette EAJA, Vennix JAM, Van Hooff P, Jongebreur W (2007) Modeling crime control in the Netherlands: insights on process. In: Sterman, JD, Oliva R, Langer RS, Rowe JI, Yanni JM (eds) Proceedings system dynamics conference. Boston, MA, USA, pp 1-26

63. Sterman JD (2000) Business dynamics: systems thinking and modelling for a complex world. McGraw-Hill, New York
64. System Dynamic Society (2013). What is SD? Available at http://www.systemdynamics.org/what-is-s/. Accessed 27 Aug, 2013
65. Ventana Systems Inc. (2010) Vensim reference manual. Ventana Systems Inc., Harvard, MA
66. Wolstenholme EF (1990) System enquiry: a system dynamics approach. Wiley, Chichester
67. Wu Z, Xu J (2013) Predicting and optimisation of energy consumption using system dynamics—fuzzy multiple objective programming in World Heritage areas. Energy 49:19–31
68. Yudken JS, Bassi JS (2009) Climate policy and energy-intensive manufacturing: the competitiveness impacts of the American Energy and Security Act of 2009. High Road Strategies and Millennium Institute, Feb 2010, Washington DC, USA. Prepared for the Environmental Defense Fund (EDF)
69. Zagonel AA (2004) Developing an interpretative dialogue for group model building. In: Proceedings of the 2004 international conference of the system dynamics society, July 25–29, Oxford, UK
70. Zhao W, Ren H, Rotter VS (2011) A system dynamics model for evaluating the alternative of type in construction and demolition waste recycling centre—the case of Chongqing, China. Resour Conserv Recycl 55:933–944

Chapter 5
System Dynamics Application to Housing Energy and Carbon Emissions in the UK

5.1 Introduction

This chapter presents the main features and structures of the model in this book. The chapter starts by discussing the details of the experts who participated in the model conceptualisation stage of the modelling process. Variables included in the model, especially the key ones showing historical trends, are important in system dynamics models. The chapter also presents the reference modes of key variables in the model as evidenced from historical data. Furthermore, the dynamic hypotheses (casual loop diagrams) including the stock and flow diagrams (model formulation) for each of the submodules in the model are described and discussed. The chapter concludes by discussing details of the variables excluded from the model.

5.2 Details of Participants in Model Conceptualisation

In applying the system dynamics approach to the modelling of energy and carbon emissions in the UK housing sector, this book used the procedure discussed in Sect. 4.5. Understandably, the research problem has been articulated and synthesised in Sect. 1.2 of Chap. 1 of this book. As such, the model conceptualisation stage involved experts and industry practitioners who are knowledgeable regarding the subject. Therefore, the main purpose of this section is to elaborate on the details of those experts who participated in the model conceptualisation stage. It is appropriate to mention that prior to this exercise, the modeller had already identified the variables relevant to the scope of the study and linked them with the use of causal diagramming, as will be seen in succeeding sections (Sects. 5.3–5.6). Each of the causal diagrams was studied in detail, and feedback loop structures within these diagrams were located and labelled appropriately. The developed causal loop

M. G. Oladokun and C. O. Aigbavboa, *Simulation-Based Analysis of Energy and Carbon Emissions in the Housing Sector*, Green Energy and Technology,
https://doi.org/10.1007/978-3-319-75346-1_5

diagrams were used to establish the effect of one variable on the others as it was hypothesised that they are likely to influence household energy consumption.

After the input from the experts, Sect. 5.5 contains the final and validated CLDs as examples used in this book. In order to validate these CLDs, the research adapted the approach used by Mohammed [11] by showing all the diagrams to each of the experts who participated in the unstructured interviews (details of the participants are shown in Table 5.1) to perform the following tasks:

- Verify the existence of the link with '1' for 'there is link' and '0' for 'there is no link'.
- Indicate the strength of the link with '3' for 'strong link', '2' for 'reasonable link', and '1' for 'weak link'.
- Verify the direction of link with '+' for 'agree the direction' and '−' for 'disagree the direction'.
- Indicate any missing link(s), using the above value.

As is evident from Table 5.1, the organisation types to which the interviewees belong are either public or private sector (50% are from the public sector, and 50% are from the private sector). Therefore, this mix strikes a balance between the public and private sector interviewees as there may likely be some differences regarding their perception of issues relating to household energy consumption. Table 5.1 further indicates that the lowest academic qualification of the interviewees is a bachelor's degree (50% of the interviewees). A total of 40% of those interviewed hold a master's degree, while the remaining 10% hold a Ph.D. degree. The

Table 5.1 Background information of experts who participated in casual diagrams validation

Category	Classification	Frequency	Percentage (%)
Organisation type	Public	5	50
	Private	5	50
	Total	10	100
Academic qualification	Bachelor's degree	5	50
	Master's degree	4	40
	Ph.D.	1	10
	Total	10	100
Years of experience in household energy-related issues	6–10	3	30
	11–15	2	20
	16–20	4	40
	21–25	1	10
	Mean = 17.5		
Experience in system dynamics modelling	No	10	100
	Yes	0	0
	Total	10	100

implication of this is that all the interviewees have the requisite academic qualifications to be presumed knowledgeable about the issues being examined by the study.

Table 5.1 indicates the years of experience of the interviewees to ensure that those interviewed have been involved in and have a deep knowledge of issues relating to household energy consumption. The results indicate that the interviewees have an average of 17.5 years of experience on issues relating to household energy. In addition, the result shows that none of the interviewees have pre-knowledge or experience in SD modelling. From the foregoing background information of experts who participated in the causal diagrams validation process, it can be concluded that the evaluation will be made by relevant and qualified experts whose inclusion in producing the final causal diagrams can be relied upon to serve as a true representation of reality in the field of study as exemplified in this book.

Having established the reliability of the interviewees' participation in the validation of the causal diagrams, interviews were conducted. Each interview began with a brief description of the research by highlighting its aim and objectives. This was then followed by an explanation of the approach adopted for the study and the expected outcome from the unstructured interview. This was necessary in order to ensure that the exercise was clear enough to the interviewees. After this, the interviewee was given the causal diagrams that were produced based on the authors' knowledge of the system under study as captured from the review of the extant literature, government reports and documents. The diagrams were appropriately explained to the interviewees. Each of the interviewees was then asked to make a review of the variables in the causal diagrams. Following this, they were asked to assess the appropriateness of the causal links for the variables included in the diagram and to suggest additional variables and links should there be any need for such. Also, they were asked to strike out some of the variables or links which they deemed unsuitable or alter the direction of the variables' links as necessary. This validation exercise therefore ensured that the views of energy experts were captured and reflected in the causal diagrams. The duration of each interview was between 45 and 75 min.

5.3 Variables of the Model

Energy and carbon emission issues are highly complex systems about which a number of decisions need to be made on a continual basis. Considering the amount of detail and information required, any attempt to model all the activities within this domain constitutes an exercise in futility. Such a model would be undesirable mainly because its complexity would obscure the dynamic nature of the parameters being observed. To this end, it is necessary to carefully select a level of aggregation in order to ensure that the model built gives all the essential parameters and policy levers required. The example in this book then combines both the top-down and bottom-up approaches (as explained in Sect. 2.5 of Chap. 2) in selecting all the variables, and as such, all the variables are aggregated at the level of policy makers in top-level management regarding housing energy and carbon emissions.

As previously highlighted, the boundary of the model needs to be carefully selected. According to Sterman [16], a model boundary chart summarises the scope of the model by listing which key variables are included endogenously, which are exogenous, and which are excluded from the model. Therefore, the model requires including all the important variables that need consideration by the policy makers together with some variables beyond the system control. To this end, the variables included in this research are extracted from the extant literature, government documents and reports as well as inputs from the knowledge elicitation process of the interviews conducted with the experts on the subject and industry practitioners as explained in Sect. 5.2 of this chapter. Therefore, Table 5.2 shows the variables included and those that are excluded from the model at this stage: this list is not exhaustive. It is necessary to mention that within the SD modelling paradigm, variables that are included in the model are divided [12] into (1) endogenous—dynamics variables that form the internal structure of the system and (2) exogenous—variables the values of which are not directly affected by the system. As given in Table 4.2 and explained in Sect. 4.6 of Chap. 4 of this book, the variables that are designated 'S' and 'F' illustrate stock and flow variables, respectively. However, all other undesignated variables represent auxiliary, data, constant, or lookup variables.

Table 5.2 Model boundary

Endogenous variables	Exogenous variables/parameters	Excluded variables
Heat losses	Total floor area	Some variables relating to occupants' behaviour, e.g.: – Occupants' social class influence – Occupants' social group influence – Occupants' cultural influence – Occupants' personal influence
Dwelling heat gain (DHG) due to cooking	Area of opening	
DHG due to no. of people	Solar flux	
DHG due to appliances less cooking	Solar transmittance factor for glazing	
DHG due to artificial lighting	Frame factor	
Average effect of solar gains	Average solar access factor	Some variables relating to dwellings physical parameters, e.g.: – Dwelling exposure – Dwelling orientation – Air changes
Total dwelling heat gains	Pi	
Natural heat transfer (F)	DHG due to water heating	
Artificial heat transfer (F)	Insulation factor	

(continued)

Table 5.2 (continued)

Endogenous variables	Exogenous variables/ parameters	Excluded variables
Dwelling internal heat (S)	Set-point temperature	Some variables relating to external environment such as: – Political uncertainties – Energy securities
Discrepancy in internal and external temperatures	Temperature conversion factor	
Dwelling internal temperature	Growth in occupants activity level	
Discrepancy in internal and set-point temperatures	External air temperature	
Humidex value	Relative humidity	
Occupants' activity level (F)	SAP rating	
Occupants' metabolic build-up (S)	Average annual gas bill	
Perceived dwelling temperature (F)	Average annual electricity bill	
Occupants' comfort (S)	Energy to carbon conversion factor	
Probability of window opening	Carbon depletion factor	
Probability of putting on clothing	Demand for cooking energy	
Effect of energy efficiency standard improvement on dwelling energy efficiency	Lighting energy demand	
Effect of fabric insulation on energy efficiency	Appliances energy demand	
Effect of combined fabric insulation and energy efficiency standard on dwelling energy efficiency	Population equilibrium time	
Effect of energy efficiency standard on cooking energy	Reproductive time	
Effect of energy efficiency standard on appliances energy	Average total fertility rate	
Effect of energy efficiency standard on lighting energy	Average life expectancy	
Effect of energy efficiency on hot water energy	Occupants' behaviour	

(continued)

Table 5.2 (continued)

Endogenous variables	Exogenous variables/ parameters	Excluded variables
Effect of energy efficiency on space heating energy		
Effect of dwelling energy efficiency on energy bills		
Effect of energy bills on energy consumption		
Climatic effects on international energy price		
Climatic effects		
Average annual energy bills		
Space heating demand		
Space heating energy rate (F)		
Space heating energy consumption (S)		
Energy to carbon conversion (F)		
Space heating carbon emissions (S)		
Carbon depletion (F)		
Rate of hot water energy usage (F)		
Hot water energy usage demand		
Hot water energy consumption (S)		
Carbon emissions due to hot water usage (S)		
Cooking energy rate (F)		
Cooking energy consumption (S)		
Carbon emissions due to cooking energy (S)		
Rate of lighting energy usage (F)		
Lighting energy consumption (S)		
Carbon emissions due to lighting energy (S)		
Appliances' energy consumption (S)		
Rate of appliances energy usage (F)		
Carbon emissions due to appliances' energy (S)		
Average annual energy consumption per household		
Total annual household energy consumption		
Average annual carbon emissions per household		
Total annual household carbon emissions		

(continued)

Table 5.2 (continued)

Endogenous variables	Exogenous variables/ parameters	Excluded variables
Population (S)		
Births (F)		
Deaths (F)		
Mortality		
Households		
Household size		

Motawa and Oladokun [12]
S stock; *F* flow

5.4 Reference Modes of Key Variables

In SD modelling, a reference mode is regarded as an important element of the modelling process. It is necessary to consider the historical behaviour of the key variables in the system under consideration and what their behaviour might be in the future [12]. Therefore, a reference mode depicts a pattern of behaviour that represents the dynamic nature of the problem in question. Consequently, it serves as a reference behaviour upon which the validity of a simulated model is assessed and determined. Sterman [16] argues that it is necessary to identify the time horizon for those variables considered to be of high importance to the problem under consideration. The main output of this research is the trend of household energy consumption and carbon emissions as affected by some key variables and how they respond to changes in some policy parameters. In order to identify the time horizon for the model in this research, it is necessary to consider the time horizon of historical data available on household energy consumption and carbon emissions as domiciled in the DECC [10], which are presented from 1970 and are updated on a yearly basis. Equally, it is of paramount importance to align to this research the time horizon of what carbon emission reductions would be in the year 2050 as stipulated in the Climate Change Act of 2008 in the UK. As a result of these considerations, the time horizon of 1970–2050 is therefore adopted for the model in this research. The following subsections present the reference modes for some of the key variables as evidenced from available historical data.

5.4.1 Population

The historical data of UK population analysed in Fig. 5.1 are based on the UK Housing Energy Fact File published by the Department of Energy and Climate Change [10]. The historical data of the total UK population suggest that the

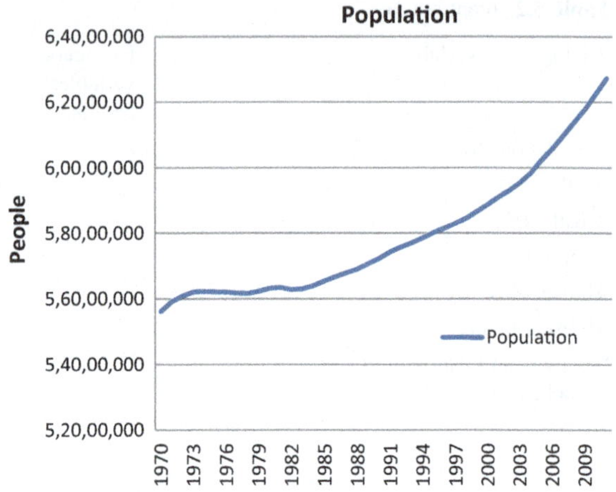

Fig. 5.1 Reference mode for population [10]

population is growing. The pattern of growth as evidenced from Fig. 5.1 reveals that between the years 1970 and 1982 the growth was minimal and not noticeable. After 1982, the growth followed a gentle slope until the year 2004. Thereafter, the slope of the growth trend has been steeper compared to that between 1982 and 2004. This trend suggests that in years to come, it is most likely that the UK population will continue to grow. The implication of this growth is profound in that population drives households and households drive household energy consumption, which in turn drives household carbon emissions. The trend therefore highlights the problem posed by the behaviour of the UK population.

5.4.2 Households

The reference mode of UK households' number is shown in Fig. 5.2. The trend indicates a growth in the number of households over the years. If this trend is sustained, it means that the number of households in the UK will continue to grow. The implication of this is possibly attendant growth in average household energy consumption, which means average household carbon emissions would witness an increase as well. This then shows that certain policies regarding the number of households could melt household energy consumption and consequently household carbon emissions. This research in this book will attempt to mimic this trend and project the trend into the future.

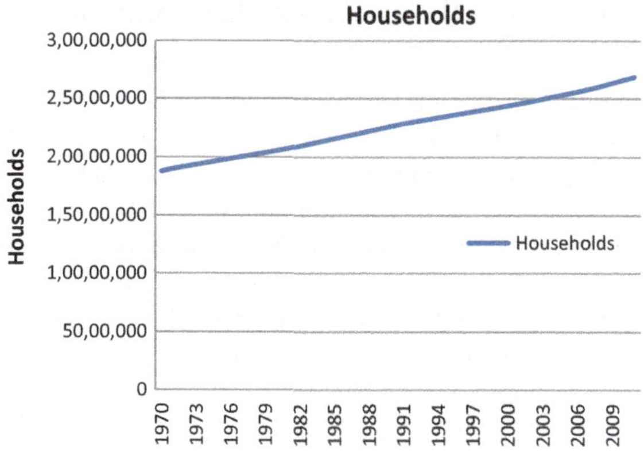

Fig. 5.2 Reference mode for households [10]

5.4.3 Space Heating Energy Consumption

Figure 5.3 shows the behaviour of household space heating energy consumption based on the available historical data. It shows that the slice of household energy consumption as a result of space heating followed an upward trajectory until 2004 when it began to decline somewhat. The growth follows a 'lumpy' trend with attendant troughs and peaks indicating the times of mild and severe weather conditions. With this reference mode, the problem associated with household space heating energy is therefore profound. The research in this book will thus attempt to simulate the trend as shown in Fig. 5.3, which will serve as the basis for validating the output of the model.

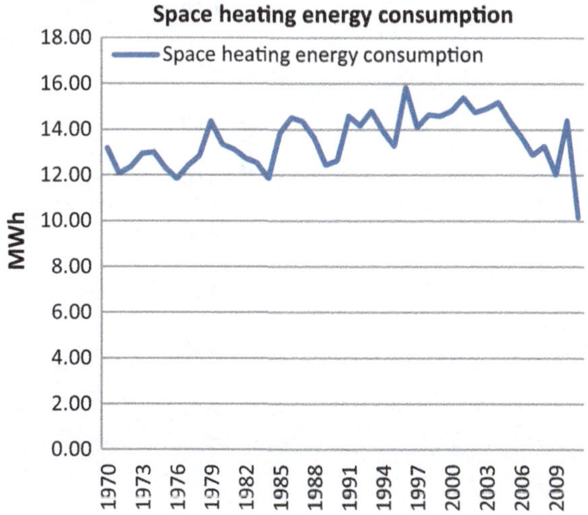

Fig. 5.3 Reference mode for household space heating energy consumption [10]

5.4.4 Hot Water Energy Consumption

In Fig. 5.4, the time series behaviour of household hot water consumption is shown. The behaviour depicted in the graph is consistent with several improvement attempts at reducing hot water energy consumption in households as the trend shows a decline in consumption since 1970. The graph indicates how hot water energy consumption in households has evolved over the years. The implication of this book is that if the trend of hot water energy is sustained, it means that less hot water energy will be consumed now than before. The insight from this reference mode will then shape the model in this research.

5.4.5 Cooking Energy Consumption

The reference mode for cooking energy use in homes as Fig. 5.5 depicts a more positive change to occupants' lifestyle as cooking energy consumption has been on the downward trend since 1970. In fact, households' cooking energy in 2009 is almost half of what it was in 1970. This shows a significant savings in cooking energy. The savings could be attributed to expansion in ready-made meals and a thriving takeaway industry. However, the trend shows a significant insight as the decline in cooking energy as from year 2000 appears to be levelling off when compared to the rate of change between 1980s and 1990s.

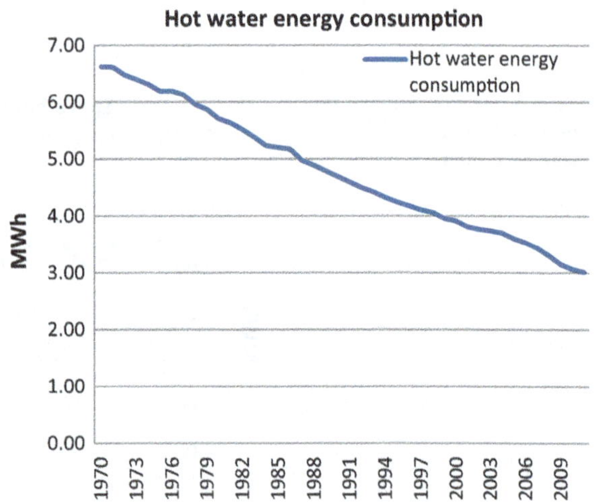

Fig. 5.4 Reference mode for household hot water energy consumption [10]

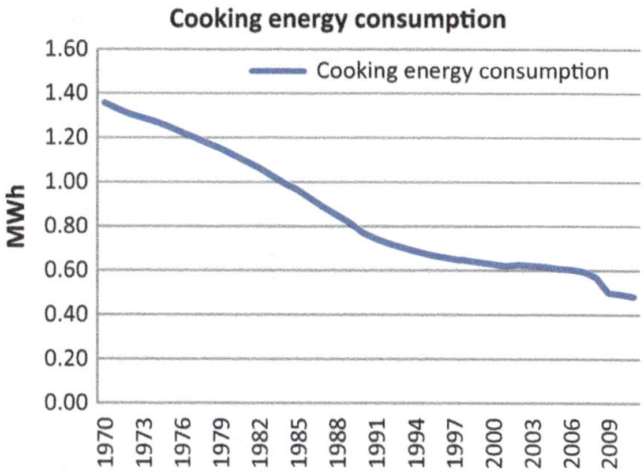

Fig. 5.5 Reference mode for household cooking energy consumption [10]

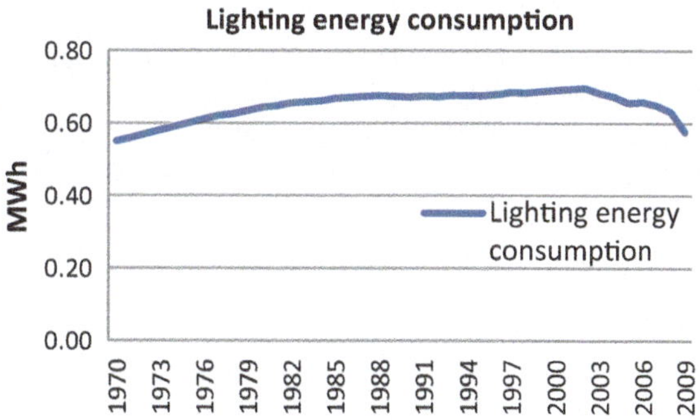

Fig. 5.6 Reference mode for household lighting energy consumption [10]

5.4.6 Lighting Energy Consumption

Undoubtedly, lighting energy has always been a fraction of household total energy. However, the insight from the reference mode as shown in Fig. 5.6 indicates that energy use for lighting was on the increase until 2002 when it began to decline. The decline reveals attempts by different schemes aimed at reducing household energy use through the use of energy-efficient lighting in homes.

5.4.7 Appliances' Energy Consumption

The reference mode for household appliances' energy consumption is shown in Fig. 5.7. The trend indicates that since 1970 appliances' energy has been on an upward trajectory.

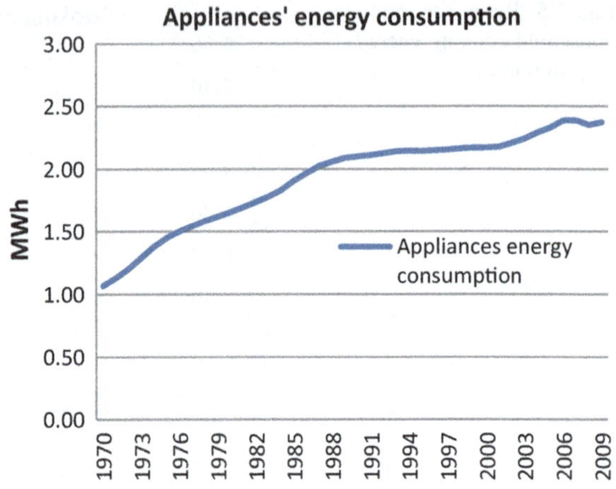

Fig. 5.7 Reference mode for household appliances energy consumption [10]

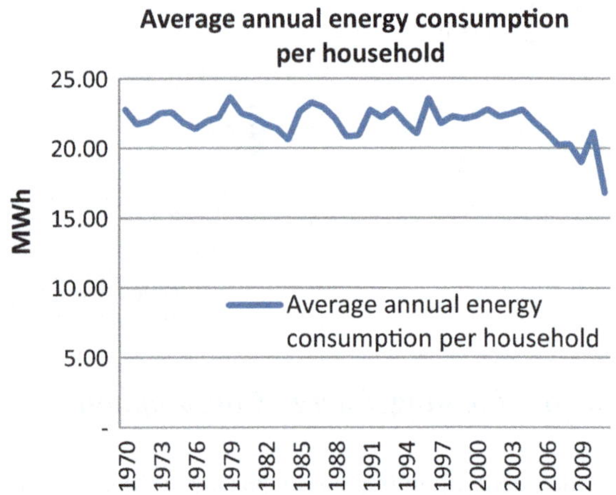

Fig. 5.8 Reference mode for average annual energy consumption per household [10]

5.4.8 *Average Annual Energy Consumption Per Household*

The pattern of behaviour of the average annual energy consumption per household as shown in Fig. 5.8 indicates that it follows the same pattern of behaviour of space heating energy consumption provided in Fig. 5.3. This further reinforces the fact that space heating energy actually shapes household energy consumption.

5.4.9 *Average Annual Carbon Emissions Per Household*

The pattern of behaviour of the average annual carbon emissions per household as depicted in Fig. 5.9 shows that household carbon emission has been on a downward

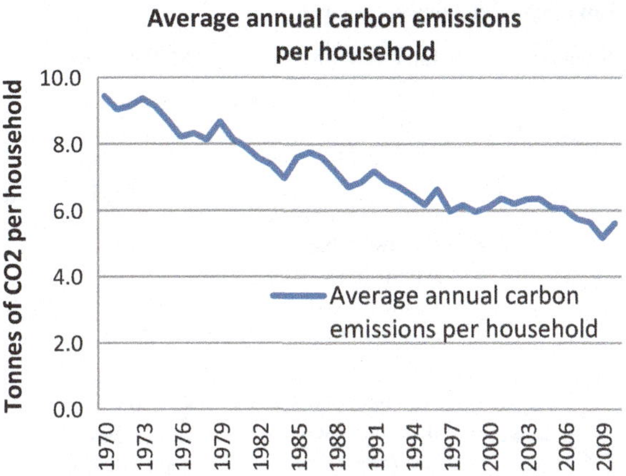

Fig. 5.9 Reference mode for average annual carbon emissions per household [10]

trend since 1970. The profile therefore follows a lumpy trend with peaks and troughs.

5.5 Model Structure I—Causal Loop Diagrams

CLDs are important tools within SD to depict the dynamic hypotheses upon which the SFDs are built. They are used to describe and understand systems under investigation [6]. Importantly, they are qualitative in nature and provide the explanation of 'cause and effect' of the underlying structure operating in a system. According to Sterman [16], CLDs are used to:

- quickly capture hypotheses about the causes of dynamics in a system;
- elicit and capture the mental models of individuals or teams; and
- communicate the important feedback believed to be responsible for a problem.

CLDs are constructed by incorporating the various variables associated with a system. Casual loops show how the variables relate with one another. That is, the relationship between any two variables is annotated by the use of an arrow connecting them. A positive relationship means an increase in arrow tail variable would cause an increase in arrow head variable and vice versa, whereas a negative relationship means an increase in arrow tail variable would cause a decrease in arrow head variable and vice versa. This relationship polarity is illustrated with symbols, interpretation, mathematics, and examples as shown in Table 5.3 based on the work of Sterman [16].

Dynamics exhibited by the system under study are achieved based on the feedback loops of the CLDs. As such, feedback loops can be positive or negative. Positive feedback loops (reinforcing loops) denote that the system increases or decreases indefinitely, whereas negative feedback loops (balancing loops) stabilise

Table 5.3 Relationship polarity

Symbol	Interpretation	Mathematics	Examples
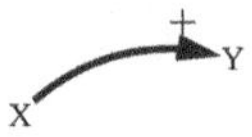	All else being equal, if X increases (decreases), then Y increases (decreases) above (below) what it would have been In the case of accumulations, X adds to Y	$\frac{\partial Y}{\partial X} > 0$ In the case of accumulations, $Y = \int_{t0}^{t} (X + \cdots)ds + Yo$	
	All else being equal, if X increases (decreases), then Y decreases (increases) below (above) what it would have been In the case of accumulations, X subtracts from Y	$\frac{\partial Y}{\partial X} < 0$ In the case of accumulations, $Y = \int_{t0}^{t} (-X + \cdots)ds + Yo$	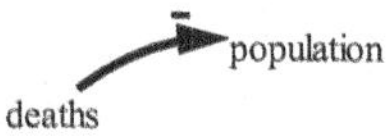

Adapted with some modifications from Sterman [12]

over time. Loops' polarity is achieved by summing up the negative polarity of each of the variables within such a loop.

The development of CLDs has its own sets of rules and guidelines to follow. Pruyt [15] provides the following guidelines for drawing CLDs:

- Make different types of CLDs for different purposes/audiences/uses (conceptualisation, loop analysis, communication, …) and at different levels of aggregation.
- Choose the right level of aggregation (but never too detailed), depending on the intended use/goal/audience.
- Iterate, use SD software to redraw your diagrams.
- Use nouns or noun phrases with a clear (positive) sense of direction as variable names. Choose variable names that, together with the causal links and polarities, enable one to easily 'read the loops'.
- Do not use or conjugate verbs in variable names. The arrows with their polarities perform the role of verbs when reading a CLD.
- Links between variables are causal and direct, not correlational nor indirect.
- Unambiguously label link polarities (split out links into different effects if polarities are ambiguous).
- Links should be drawn and interpreted under the *ceteris paribus* assumption, i.e. that everything else remains the same.
- Links are relative: they indicate the value of the variable will be above or below what it would have been without the effect.

- Include the goals of goal-seeking loops explicitly.
- Distinguish between actual versus perceived conditions.
- Trace and label loop polarities unambiguously, and name loops so that they immediately convey their role, and so that the names can be used in texts and presentations.
- Indicate important delays on causal links.
- Use curved lines, make important loops circular, and minimise crossed lines.
- For communication purposes, do not use too large a diagram with too many loops.
- Do not try to make comprehensive or final CLDs: they will never be. They are either conceptualisation tools or tools to communicate (specific) messages.
- Draw CLDs from different angles and perspectives and at different levels of aggregation.
- It is useful to integrate points of view in one and the same CLD, unless the points of view represent fundamentally different or irreconcilable world views. Then, different CLDs need to be drawn.

In this book, CLDs are constructed for each of the modules in the model. All of modules seamlessly work together in an interlinked manner. CLDs for each of the modules are presented as follows:

5.5.1 Population/Household Module Causal Loop Diagram

The causal loop of variables purported to be driving population and households in the study area of the example in this book is shown in Fig. 5.10. Some loops were created in the CLD to depict the kind of dynamics found in this module. For example, there is a combination of both positive and negative loops that give rise to the kind of behaviour expected in this sector of the model. According to Motawa and Oladokun [12], the positive loop can be interpreted as '*…the more the people, the more the births there will be; the more births there are, the more people there will be. Alternatively, this could be read as the fewer people there are, the fewer births there will be; the fewer births there are, the fewer people there will be*'. Following this interpretation, the negative loops can be interpreted the same way as well.

The model postulates that the number of births is based on the number of deaths, reproductive time, population equilibrium time, average total fertility rate, and of course population as well [8, 9]. Referring to the positive or reinforcing loop as shown in Fig. 5.10, the feedback loop could be interpreted as an increase or decrease in population that will cause an appropriate increase or decrease in the number of births, respectively. In the negative or balancing feedback loop as shown in Fig. 5.10, however, the loops will serve as stabilising factor for the entire CLD of population or household module. For example, the continual growth or decline will be balanced out and stabilised. It needs to be mentioned that population drives

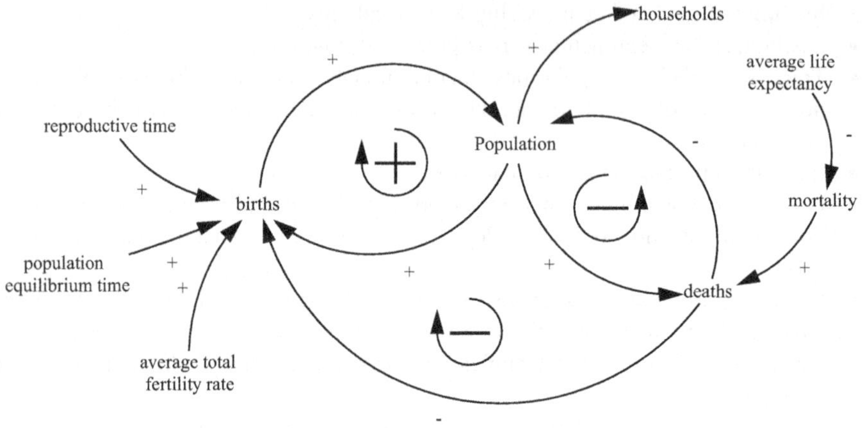

Fig. 5.10 Population/household module CLD

deaths and vice versa. Also, average life expectancy drives mortality, and mortality in turn drives deaths.

The interaction of different variables in this module as shown in Fig. 5.10 is expected to generate interesting output behaviour considering the fact that the module consists of multiloops. As such, some of the loops are expected to be more dominating than the others. If that is the case, a scenario may suggest growth in population should the reinforcing loop be the one dominating the others, though the increment value may vary. On the other hand, should the balancing feedback loop be the one dominating, this may generate another interesting behaviour in that the number of deaths will continue to grow, while there will be a corresponding decrease in population up to the point that the entire population will face extinction. Once there is population extinction, then the number of households will flatten and approach zero.

5.5.2 Dwelling Internal Heat Module Causal Loop Diagram

This module as shown in Fig. 5.11 used building physics to show the interaction of the occupants with the dwelling internal heat. The CLD in Fig. 5.11 indicates the complexity involved in the thermodynamics of dwellings' internal heat. For example, the CLD suggests that the causal structure of the variables is purported to be influencing the internal heat of dwellings. Within this dwelling internal heat system, the CLD indicates that the dwelling internal heat is driven by a combination of variables relating to the natural and artificial heat transfers and dwelling heat gains. This means that should a change occur in the internal and external temperatures within the dwelling, there will be a temperature gap that revolves round the set-point temperature based on the response of the occupants to their thermal environment. Significantly, the temperature set-point and dwelling internal temperature moderate the artificial heat transfer, and these in turn moderate the dwelling internal heat within this system.

Figure 5.11 further suggests that dwelling heat gains are aggregated from many sources. These sources include those from solar gains, artificial lighting, cooking, water usage, appliances use, and the number of people in the dwelling. However, the lack of proper airtightness within the dwelling may result in infiltrations, which may cause heat losses from the dwelling and consequently reduce the dwelling heat gains. The resultant rate of dwelling heat gains will drive the dwelling internal heat. Consequently, the dwelling internal heat will influence the dwelling internal temperature, which will accordingly influence the perceived dwelling temperature of the occupants as driven by some other environmental factors.

As shown in Fig. 5.11, two different feedback loops were built for the module: one is reinforcing and positive, while the other is a balancing or negative feedback loop. As previously discussed under Sect. 5.5.1, the reinforcing loop suggests either growth or decrease in the value of dwelling internal heat as the case may be, while the balancing feedback loop will stabilise this system over time as regulated by the effect of artificial heat transfer in dwelling. This interaction is expected to generate an interesting insight based on the behaviour of the output. The kind of behaviour that is expected could be in the form of S-shaped behaviour. This may be due to the fact that the behaviour is being moderated by the combination of the reinforcing feedback loop with that of the balancing feedback loop. However, the nonlinear

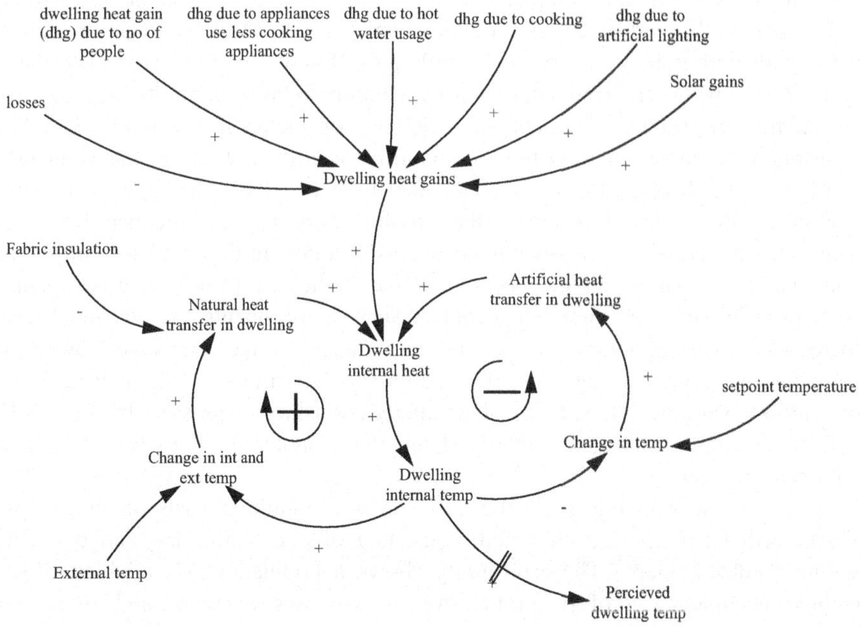

Fig. 5.11 Dwelling internal heat module CLD

behaviour could be shifted and regulated according to loop predominance as discussed in Sect. 5.5.1, and hence, the behaviour may not be S-shaped as postulated above.

5.5.3 *Occupants' Thermal Comfort Module Causal Loop Diagram*

In terms of occupants' thermal comfort, the CIBSE [4] separates the thermal environment of occupants in dwellings into three distinct categories, namely thermal comfort, thermal discomfort, and thermal stress. According to the International Standard Organisation (ISO) 7730 [9], the American Society of Heating, Refrigerating and Air Conditioning Engineers (ASHRAE) [1], and the CIBSE [4], thermal comfort is defined as '*…that condition of mind which expresses satisfaction with the thermal environment*'. This is the condition that indicates whether a person is feeling either too hot or too cold. Conversely, the CIBSE [4, 5] describes thermal discomfort as the condition when people start to feel uncomfortable, but without any unwell conditions. Likewise, the CIBSE [4] reports that the thermal stress of occupants illustrates uncomfortable conditions for occupants with attendant potential to result in a detrimental medical status to the occupants.

From the foregoing, it is important to note that the occupants' thermal environment is something that can neither be expressed in degrees nor satisfactorily determined by acceptable temperature ranges. As such, it is a matter of personal experience as dictated by many variables, which are most likely to differ among individuals within the same space. According to Motawa and Oladokun [12], these variables can be (1) environmental—air temperature, relative humidity, air velocity, or radiant temperature; (2) personal—clothing or metabolic heat; or (3) other contributing variables—access to food and drink, acclimatisation, or state of health.

Figure 5.12 displays the CLD conceptual model of the module of occupants' thermal comfort. The CLD depicts the variables purported to influence the occupants' thermal comfort as shown as occupants' comfort in Fig. 5.12 for simplicity sake. The model shows an array of variables hypothesised to influence occupants' comfort in this module, including variables such as 'occupants' activity level' and 'perceived dwelling temperature'. The importance of the 'perceived dwelling temperature' variable to the occupants' comfort is worth mentioning as it received five inflow variables (Fig. 5.12). The interrelationships expressed by the CLD indicate the kind of complexity involved, and the variables are likely to interact in a nonlinear manner.

As can be seen in Fig. 5.12, the CLD for the module consists of multiloops. Three feedback loops are identified, consisting of one reinforcing and two balancing feedback loops. The reinforcing shows a population of variables which include occupants' comfort, probability of windows opening, and perceived dwelling temperature. In the first balancing feedback loop, the variables that are

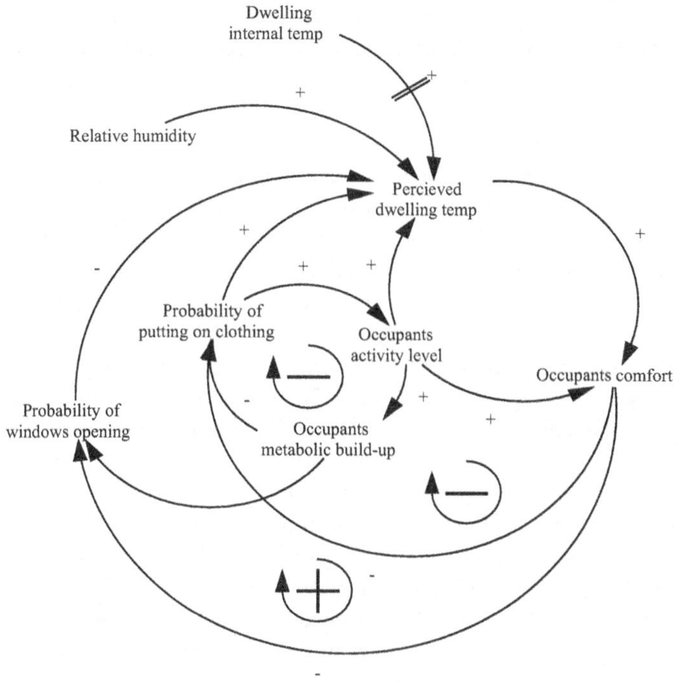

Fig. 5.12 Occupants' thermal comfort module CLD

involved include an interaction between occupants' comfort, probability of putting on clothing, and perceived dwelling temperature. The second balancing feedback loop, however, has occupants' activity level, occupants' metabolic build-up, and probability of putting on clothing as the variables. As discussed for other modules in Sects. 5.5.1 and 5.5.2 of this book, this module is expected to generate interesting insights into the behaviour of the main variables here as predominately moderated by the multi feedback loops as shown within the CLD.

5.5.4 Climatic-Economic-Energy Efficiency Interaction Module Causal Loop Diagram

This module indicates how the nexus of climatic, economic, and energy efficiency issues moderate energy consumption in dwellings. The variables within these three areas can interrelate and seamlessly work together to illustrate the kind of complexity involved. The variables considered are not limited to climatic effects as a result of adverse weather conditions. However, the influence of international fuel prices is seen as one of the key variables that can activate the policy of the government about issues involving energy tax and matters that relate to alternative energy sources and the costs attached to these energy sources. Also, this may influence the policy of the government about energy efficiency of dwellings, fabric insulation, and the income of the occupants. Fuel poverty is therefore likely to have an influence on the health and well-being of the occupants.

Considering the simulation of the above-mentioned variables, the experts that were interviewed during the course of the study being reported in this book suggested that not all the variables can be included. Therefore, Fig. 5.13 indicates the variables that are included in this module. Two views are used to illustrate the system within the dwelling energy. The first is the fabric insulation and how this can influence energy efficiency. The second perspective used the government's Standard Assessment Procedure (SAP) rating of dwellings to indicate its influence on the energy efficiency of the dwellings. The causal model as shown in Fig. 5.13 assumed that the effect of energy efficiency standard improvement on dwelling energy efficiency has the capability of improving energy efficiency standard of cooking energy and that of lighting and appliances energy. The model presumed that the government's SAP rating of dwellings [2] is made up of those energy efficiency standards. The causal diagram as shown in Fig. 5.13 further hypothesised that the combined effects of fabric insulation and energy efficiency standard of dwellings are likely to grow the effect of dwelling energy efficiency on energy bills, which in turn is likely to influence energy for space heating and hot water usage.

Additionally, the casual model hypothesised three inflow variables for 'the effect of energy bills on energy consumption'. According to Fig. 5.13, the three inflow variables are 'average annual energy bills' in terms of gas and electricity bills, the 'effect of dwelling energy efficiency on energy bills', and the 'effect of unfavourable climatic effects on energy prices'. These three variables combined give their influence on the 'effect of energy bills on energy consumption'. Furthermore, the CLD suggests that the climate change due to the accrual of carbon emissions may activate unfavourable climatic effects with an assumption that there will be a delay for this process to occur. Importantly, feedback loops cannot be isolated from the CLD presented in Fig. 5.13. However, the loops will definitely show up when this module interrelates with other modules in the model.

5.5.5 Causal Loop Diagram for the Combined Household Energy Consumption and Carbon Emission Modules

Two different modules are represented in this section, namely the household energy consumption module and carbon emission module. The modules were combined in order to show the complex interrelationships among the variables of the two modules. The causal diagram for the modules is shown in Fig. 5.14. The concept from this CLD is based on the fact that the household energy consumption drives the carbon emissions. It is noteworthy that both household energy and carbon

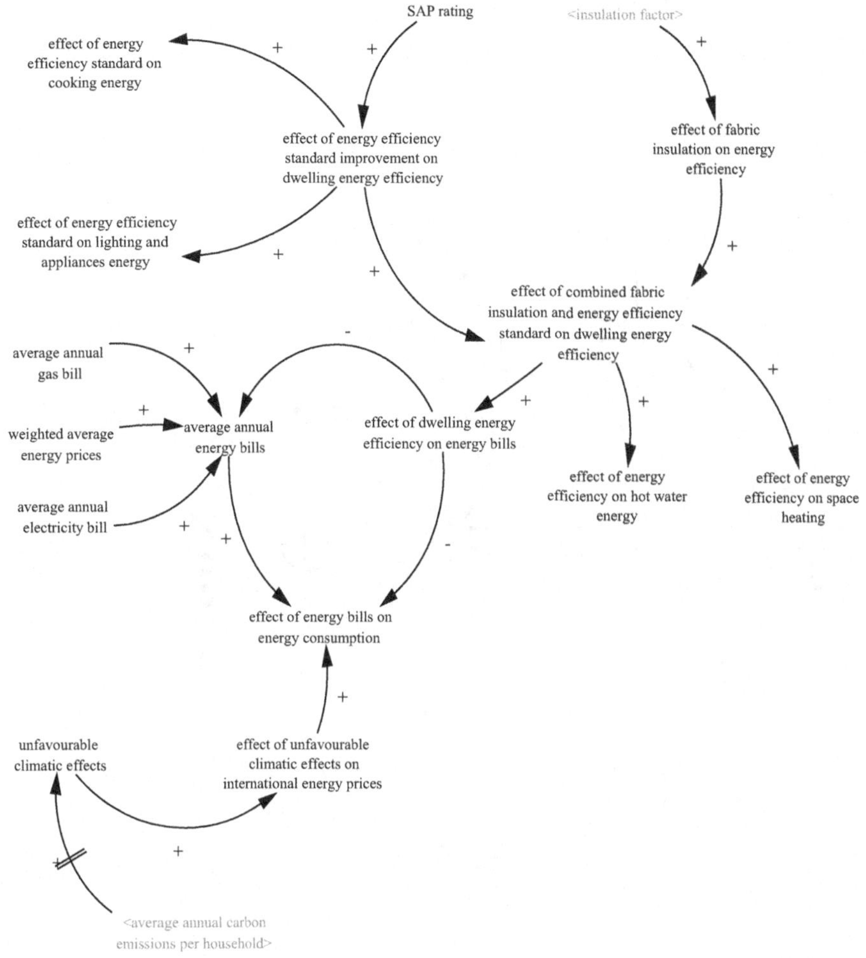

Fig. 5.13 Climatic-economic-energy efficiency module CLD

emissions are decomposed into five different energy end-uses which include space heating, hot water, appliances, cooking, and lighting. The causal model structure is then tailored towards these five end-uses of energy. However, household energy end-uses for space heating, hot water, and appliances energy are represented in Fig. 5.14 for clarification.

To serve as an example, the loops involving space heating energy consumption will be traced and discussed accordingly as shown in Fig. 5.14. This reflects that there are a number of variables driving the rate of space heating by occupants. Evidently, those variables are fully covered somewhere else [8] to include dwelling internal temperature and set-point temperature from the dwelling internal heat module; occupants' comfort (from the occupants' thermal comfort module) driving

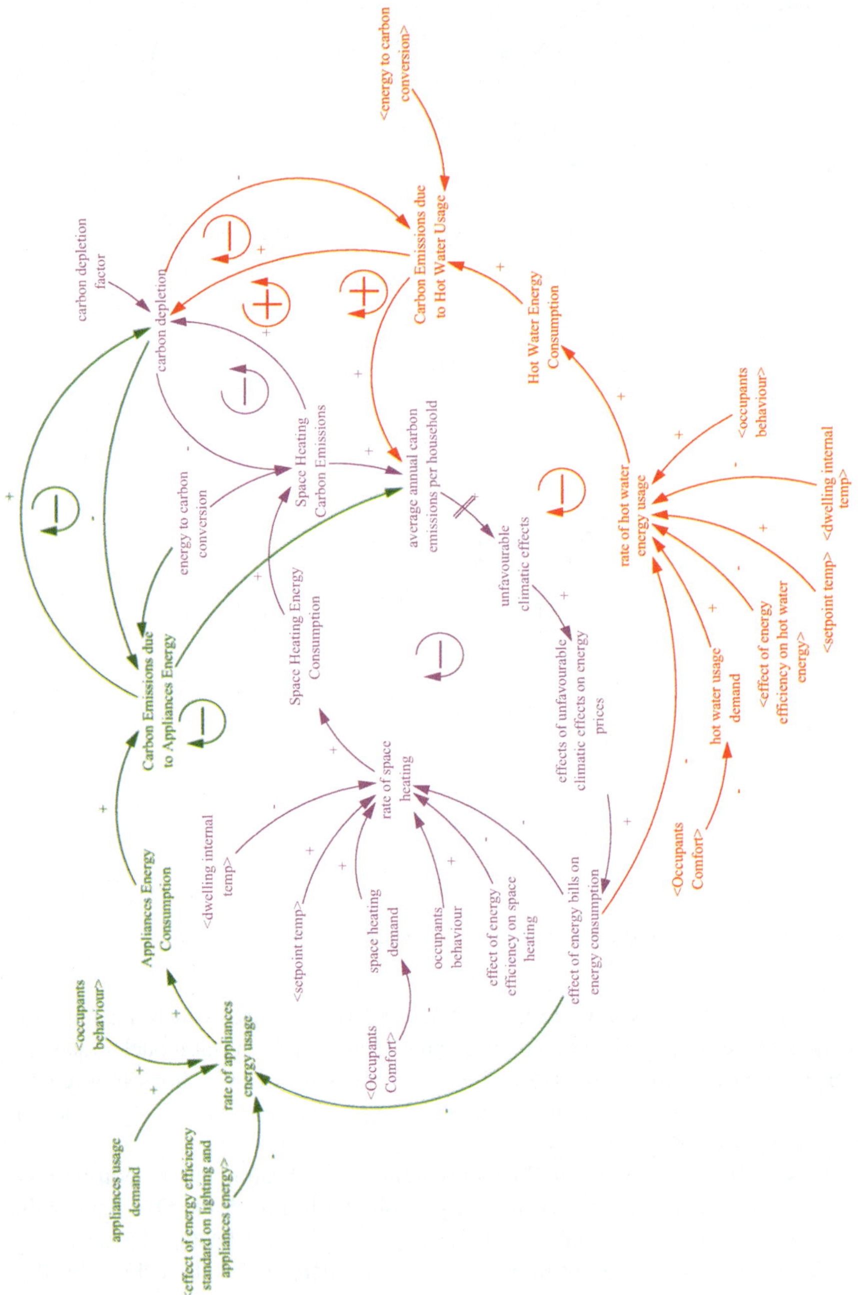

Fig. 5.14 CLD for household energy consumption and CO_2 emissions

space heating demand, and space heating demand in turn propelling the rate of space heating; effect of energy efficiency on space heating, and effect of energy bills on energy consumption (from the climatic-economic-energy efficiency interaction module); and occupants' behaviour.

As indicated in Fig. 5.14, the causal model indicates that the rate of space heating drives space heating energy consumption, which in turn drives the space heating carbon emissions. The accumulation of carbon emissions from different household energy end-uses gives the average annual carbon emission per household. This is in turn presumed to reinforce the unfavourable climatic effects' variable, though with a delay before this can occur. Accordingly, there was an assumption that some of the emitted carbon is likely to be depleted from the atmosphere by different means such as absorption by plants and will therefore not necessarily cause unfavourable climatic conditions. To this end, the model structure introduced the carbon depletion factor to control this as shown in Fig. 5.14. Based on the way the feedback loop around the space heating energy consumption is explained, other household energy end-uses can be explained and discussed.

As shown in Fig. 5.14, the CLD for the two modules has both balancing and reinforcing feedback loops. As such, the anticipated behaviour is expected to generate nonlinear outcomes as explained for the preceding modules. However, loops' predominance may moderate the final insights from the modules.

5.6 Model Structure II—Stock and Flow Diagrams

The dynamic hypotheses (CLDs) are undoubtedly useful in many situations. They capture the mental models of interdependencies and feedback processes of a given modelling exercise. However, there are a number of limitations of which their inability to capture the stock and flow structure of the systems is more prominent [12]. This then shows that there is a need for SFDs. This stage of model building is underpinned by the variables and parameters as developed by in the CLDs in Sect. 5.5. These are then transformed into SFDs. The SFDs differentiate the model parameters into the controlling 'flow', acting as regulators and 'stock' where accumulations take place. Accordingly, accumulations characterise the state of the system and generate information upon which decisions and actions are based. Stocks propel the system and give them the necessary memory that is required. Additionally, stocks create delays by accumulating the difference between the inflow to a process and its outflow [12]. Stocks are known for being the source of disequilibrium dynamics in systems when the rates of flow are decoupled. Within the SFDs, all the variables are supplied with requisite equations and algorithms with data imported into the parameters in the model. This is necessary in order to prepare for simulation. The model is built and implemented using Vensim software. The following subsections describe the SFDs for each of the modules in the model.

5.6.1 SFD for Population/Household Module

In a bid to estimate energy consumption and carbon emissions per household, it is expedient to model the number of households in the UK. Based on the regression analysis performed on historical data of the UK population and households, a significant relationship was found between the two. Hence, the model of the population was used to estimate the number of households in the UK. Population is influenced by a number of variables. In this model, a limited number of variables are considered, namely birth and death rates, reproductive time, population equilibrium time, average total fertility rate, mortality, and average life expectancy as shown in Fig. 5.15.

Population is modelled as 'stock'. This is being controlled by an inflow (births) and an outflow (deaths). 'Births' is influenced by 'reproductive time', 'population equilibrium time', 'average total fertility rate', and 'deaths'. On the other hand, 'deaths' is determined primarily by 'mortality' rate. 'Mortality' is generated based on 'average life expectancy' and the 'mortality lookup' profile. 'Mortality lookup' is one of the variables in the model that is based on the qualitative data of relying on experts' judgement as well as information from the literature. Relating different variables together in this sector of the model involves generating a set of equations by employing both regression analysis and tools within the system dynamics software (Vensim). For example, Eq. 5.1 shows the Vensim interpretation of 'population' acting as the stock with 'births' as inflow (rate) and 'deaths' as outflow (rate). This equation is automatically generated by the software. However, the equation for 'births' (see Eq. 5.2) was developed based on a little adjustment to the

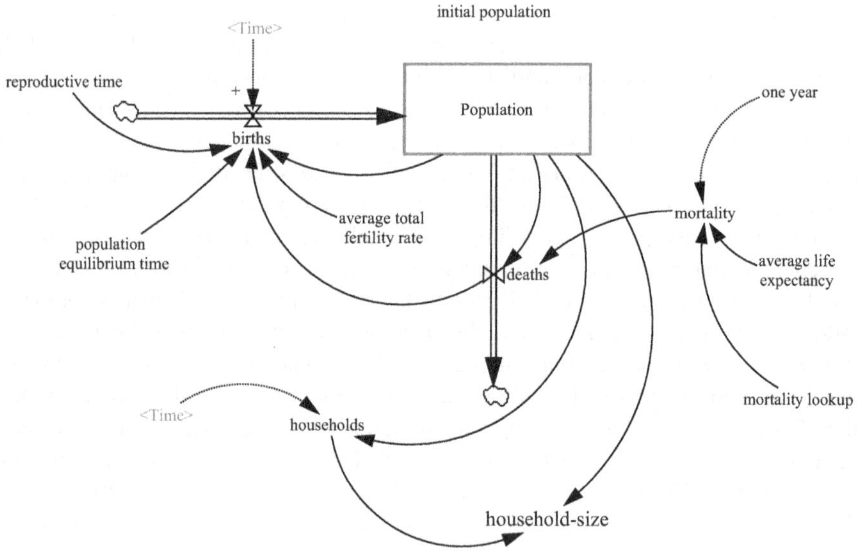

Fig. 5.15 SFD for population/household module

similar equation developed by Forrester [8] for his World Dynamics model. The adjustment made was mainly to reflect the number of years from the last year of data and the last year of model run.

Additionally, 'mortality' (see Eq. 5.4) is determined based on the profile of the mortality rate that is qualitatively captured as 'mortality lookup' according to available data from the Office of National Statistics as shown in Fig. 5.16. 'Households' is determined based on regression analysis as shown in Table 5.4. The main data source for the development of this algorithm (population and household) is shown in Table 5.5. Therefore, Eqs. (5.1)–(5.6) give the major equations developed for this module. It needs to be further stated that the model is run from the year 1970–2050 and the initial population corresponds to the population in the year 1970. The complete set of model algorithms can be found in Appendix A of this book.

$$\text{Population}(t) = \text{INTEGRAL}[\text{births}-\text{deaths}, \text{population}(t_0)] \tag{5.1}$$

$$\text{births} = \begin{cases} \text{IF THEN ELE(Time} = \text{population equilibrium time, deaths,} \\ \text{IF THEN ELSE(Time} \leq 2011, \text{average total fertility rate} * \text{Population} \\ \qquad * 0.08/\text{reproductive time,} \\ \text{FORECAST(average total fertility rate} * \text{Population} * 0.08/\text{reproductive time}, 39, 100))) \end{cases} \tag{5.2}$$

$$\text{deaths} = \text{Population} * \text{mortality} \tag{5.3}$$

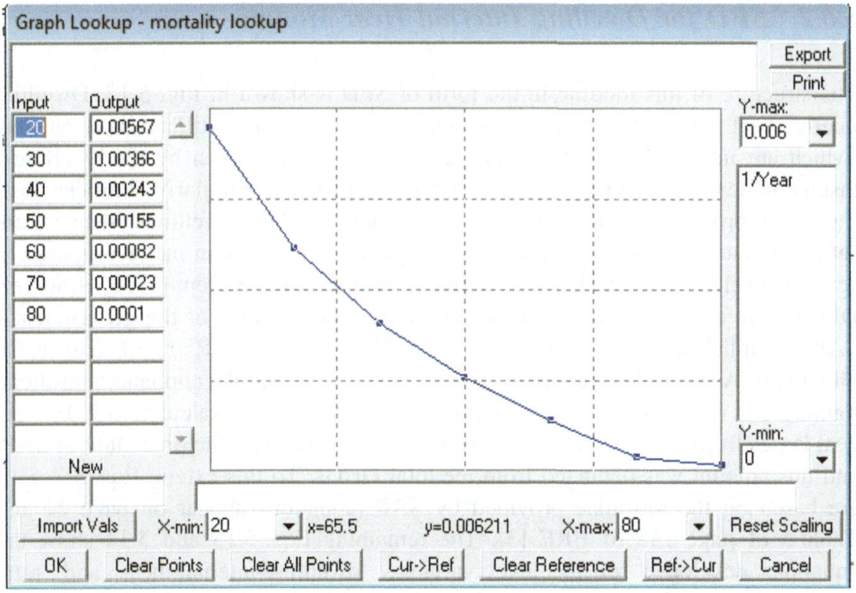

Fig. 5.16 Mortality lookup

Table 5.4 Relationship developed for households from regression analysis

Model		Unstandardized coefficients		Standardized coefficients	t	Sig.
		B	Std. error	Beta		
1	(Constant)	−3.436E8	4,538,500.746		−75.712	0.000
	Population	0.067	0.017	0.056	3.886	0.000
	Time	182,057.612	2746.898	0.947	66.278	0.000

Dependent variable: Households

Table 5.5 Sample data for households and population

Variable	Unit of measurement	Minimum	Maximum	Mean	Standard error	Standard deviation
Households	Households	18,791,000	26,863,000	2.2664E7	3.63850E5	2.35802E6
Population	People	55,632,000	62,736,000	5.7931E7	3.00346E5	1.94646E6

$$\text{mortality} = \text{mortality lookup(average life expectancy/one year)} \tag{5.4}$$

$$\text{households} = -3.436\text{e}008 + 182058 * \text{Time} + 0.067 * \text{Population} \tag{5.5}$$

$$\text{household size} = \text{Population/households} \tag{5.6}$$

5.6.2 *SFD for Dwelling Internal Heat Module*

The structure of this module in the form of SFD is shown in Fig. 5.17. Dwelling internal heat is modelled as an accumulation of natural and artificial heat transfers (which are modelled as flows). Natural heat transfer is driven by the dwellings' insulation level and internal and external temperature. Similarly, artificial heat transfer is propelled by total dwelling heat gains (DHGs), dwellings' temperature set-point, and internal temperature. The temperature conversion factor was used to convert dwelling internal heat measured in Watts to degree centigrade. However, DHGs were estimated based on the procedure and formulae of the government's SAP as published by the Building Research Establishment (BRE) on behalf of the DECC [3]. As such, DHGs due to cooking, number of people, appliances, artificial lighting, hot water, and solar gain effects were included in the calculation. Likewise, heat losses from the dwelling fabric were estimated based on the procedure of SAP, and this amount was deducted from the total DHGs. To this extent, Eqs. 5.7–5.12 are based on the formulae provided by SAP (equations appear on page 22 and Table 5 of page 145 of BRE [3]. The remaining Eqs. 5.13 and 5.14 show the equations developed for the level variables 'dwelling internal heat' and 'total dwelling heat gains', respectively. The only data driving this module are 'external air temperature', which is the yearly average for the UK as summarily shown in Table 5.6.

$$\text{heat losses} = -40 * \text{household size} \tag{5.7}$$

$$\text{dhg due to cooking} = 35 + (7 * \text{household size}) \tag{5.8}$$

$$\text{dhg due to people} = 60 * \text{household size} \tag{5.9}$$

$$\text{dhg due to appliances less cooking} = (207.8 * (\text{total floor area} * \text{household size}) * \text{EXP}(0.4714)) * (1 + 0.157 * \text{COS}(2 * pi * (\text{Time-}1.178))) * 1000/60 \tag{5.10}$$

$$\text{dhg due to artificial lighting} = (59.73 * (\text{total floor area house hold size})^\wedge 0.4714) * 0.96^\wedge 2 * (1 + 0.5 * \text{COS}(2 * \text{pi} * (\text{Time-}0.2))) * 0.85 * 1000/(24 * 30 * 12) \tag{5.11}$$

$$\text{average effect of solar gains} = 0.9 * \text{area of opening} * \text{frame factor} * \text{average solar access factor} * \text{solar flux} * \text{solar transmittance factor for glazing} \tag{5.12}$$

$$\text{Dwelling internal heat}(t) = \text{INTEGRAL}[\text{natural heat transfer} + \text{artificial heat transfer}, \text{dwelling internal heat}(t_0)] \tag{5.13}$$

$$\text{Total dwelling heat gains} = (\text{DHG due to appliances less cooking} + \text{DHG due to artificial lighting} + \text{DHG due to cooking} + \text{DHG due to no. of people} + \text{DHG due to water heating} + \text{average effect of solar gains} - \text{heat losses}) \tag{5.14}$$

5.6.3 SFD for Occupants' Thermal Comfort Module

There are a great number of techniques for estimating likely thermal comfort, including effective temperature, equivalent temperature, wet bulb globe temperature, resultant temperature, among others, and charts exist showing predicted comfort zones within ranges of conditions. However, the ISO 7730 [8] suggests thermal comfort can be expressed in terms of predicted mean vote (PMV) and percentage people dissatisfied (PPD). These were developed by Prof. Ole Fanger [7] by using the principles of heat balance equations and empirical studies regarding the skin temperature in order to define thermal comfort. In line with the PMV and PPD, the Chartered Institution of Building Services Engineers (CIBSE) [4, 5]

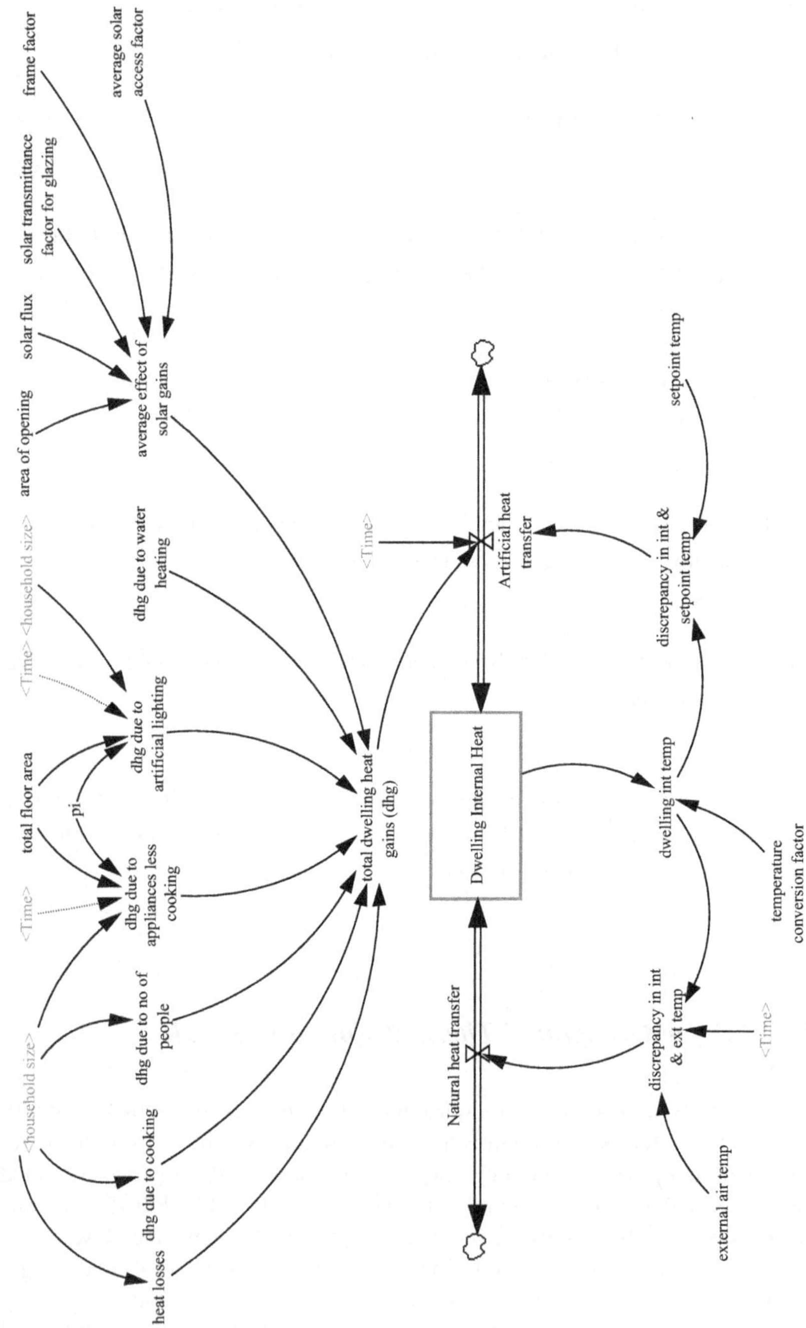

Fig. 5.17 SFD for dwelling internal heat module

Table 5.6 Sample data for external air temperature

Variable	Unit of measurement	Minimum	Maximum	Mean	Standard error	Standard deviation
External air temperature	Degree centigrade	8.06	10.80	9.78	0.10	0.69

Palmer and Cooper [14]

recommended comfort criteria for specific applications in certain areas of the dwellings in terms of temperature, occupants' activity, and clothing levels. For example, the guide [5] stipulates a winter operating temperature of 17–19 °C, activity of 0.9 met., and clothing level of 2.5 clo., for bedrooms.

In building a SFD for this module, the criteria as set out by CIBSE [5] were employed and this is presented in Fig. 5.18. Both the 'occupants' comfort' and 'occupants' metabolic build-up' were modelled as stock based on Eqs. (5.15) and (5.16). Accumulation of 'occupants' comfort' stock, for example, is driven by the 'perceived dwelling temperature' (inflow). This in turn depends on a number of factors such as humidex value, clothing, windows opening within the dwelling as well as occupants' metabolic build-up. 'Humidex value' was modelled from the dwelling internal temperature and relative humidity based on Fig. 5.19 which shows different ranges of humidex value for different degrees of comfort by qualitatively mimicking it with the use of lookups within the model. As shown in Fig. 5.18, the 'probability of window opening' and 'probability of putting on clothing' by occupants were determined qualitatively using lookups as shown in Figs. 5.20 and 5.21, respectively, according to the mental data collected from the interviewees at the model conceptualisation stage. The main data driving this module are from relative humidity (summary shown in Table 5.7), which is externally sourced and internally generated data from the previous module (i.e. internal dwelling temperature). Examples of equations developed for this module are shown in Eqs. (5.17)–(5.19).

$$\text{Occupants' comfort}(t) = \text{INTEGRAL}[\text{perceived dwelling internal temperature}, \text{occupants comfort}(t_0)] \tag{5.15}$$

$$\text{Occupants metabolic build-up}(t) = \text{INTEGRAL}[\text{occupants activity level} + \text{perceived dwelling internal temperature occupants metabolic build-up}(t_0)] \tag{5.16}$$

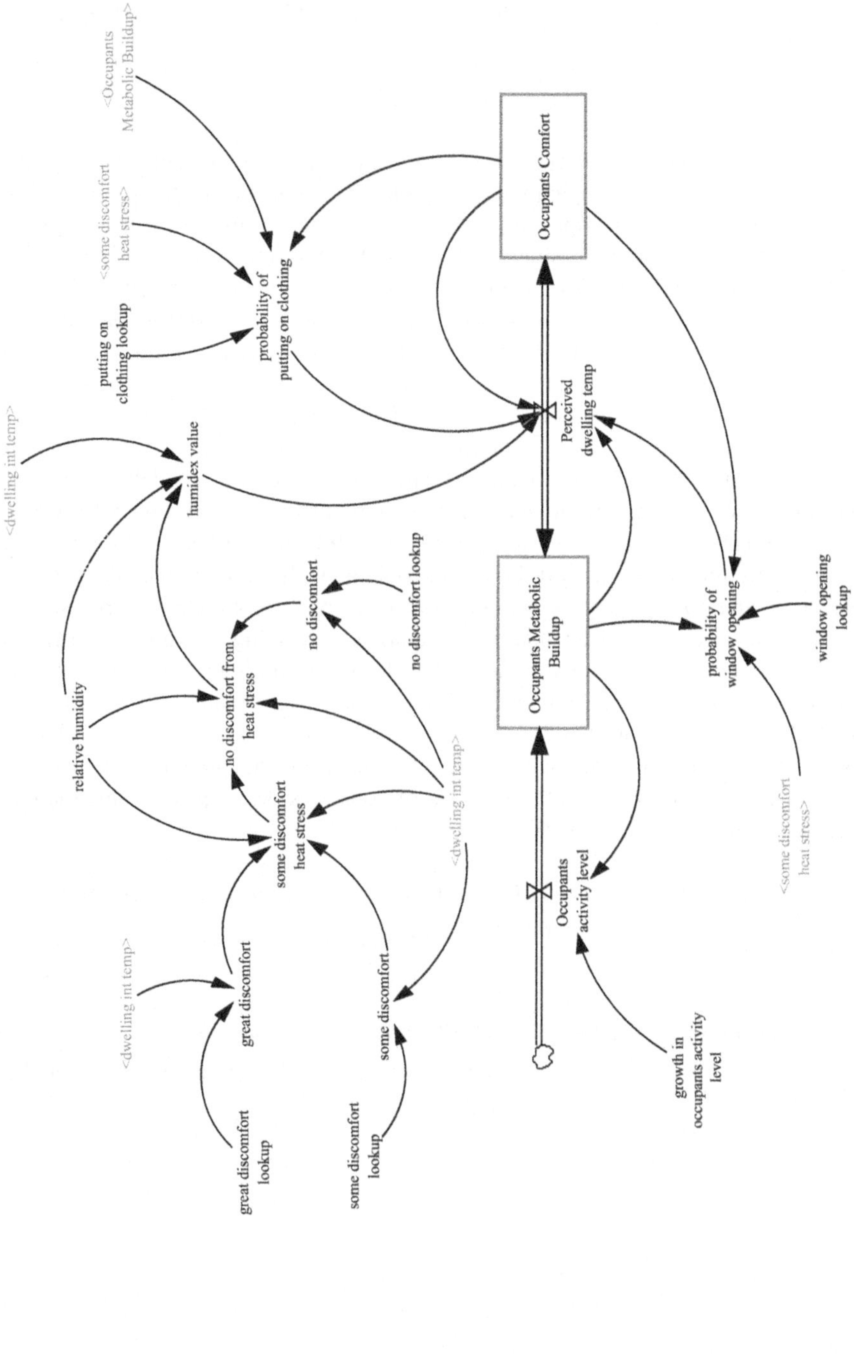

Fig. 5.18 SFD for occupants' thermal comfort module

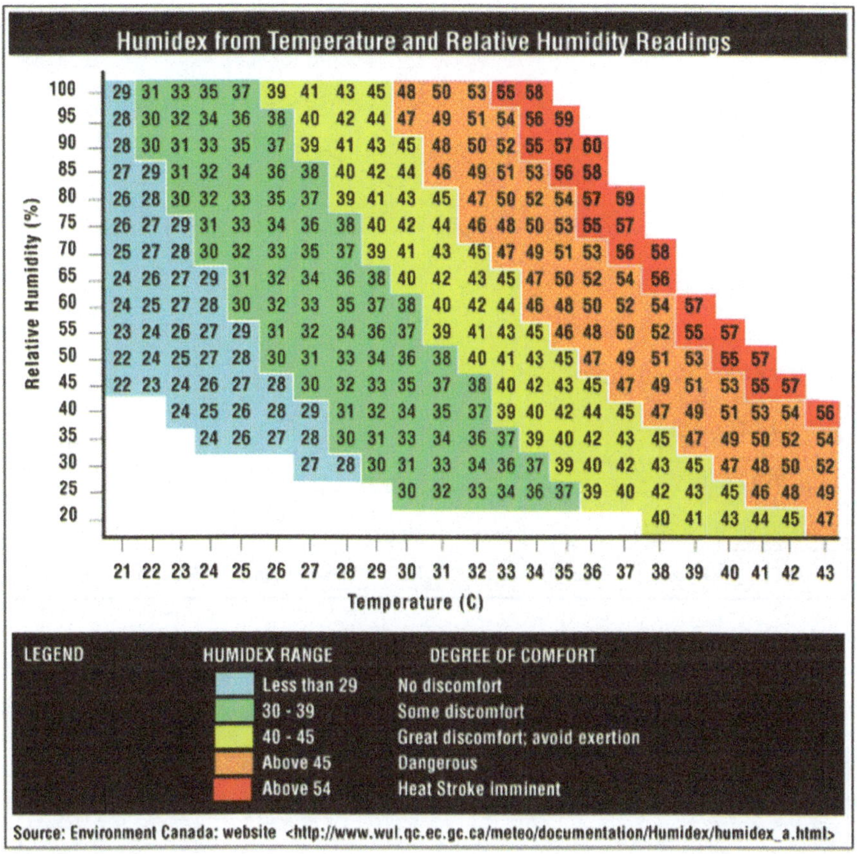

Fig. 5.19 Humidex chart

$$\text{humidex value} = \left\{ \begin{array}{c} \text{IF THEN ELSE(dwelling internal temperature} < 21 : \text{AND} : \\ \text{relative humidity} < 45, \\ \text{dwelling internal temperature, no discomfort from heat stress)} \end{array} \right. \tag{5.17}$$

$$\text{no discomfort from heat stress} = \left\{ \begin{array}{c} \text{IF THEN ELSE(dwelling internal temperature} < 30 : \text{AND} : \\ \text{relative humidity} > 25, \\ \text{no discomfort from heat stress, some discomfort from heat stress)} \end{array} \right. \tag{5.18}$$

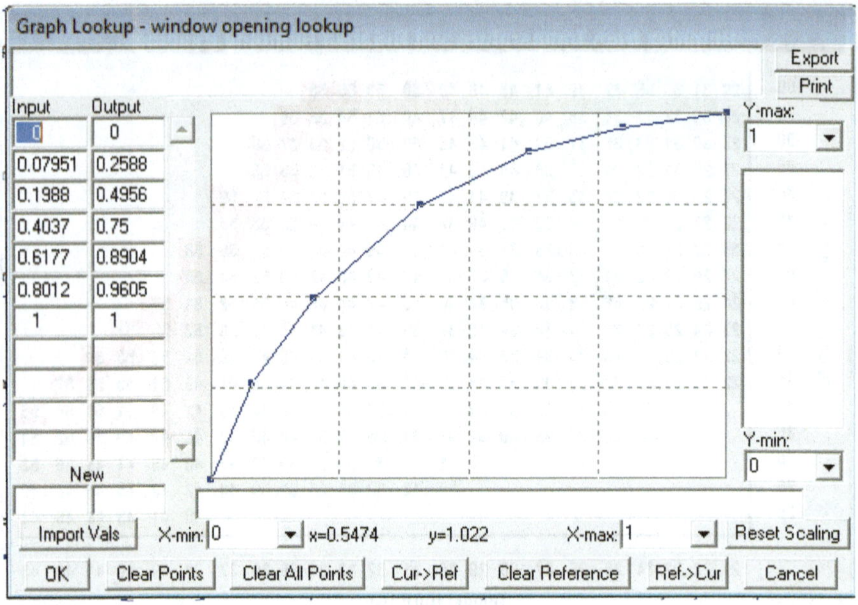

Fig. 5.20 Window opening lookup

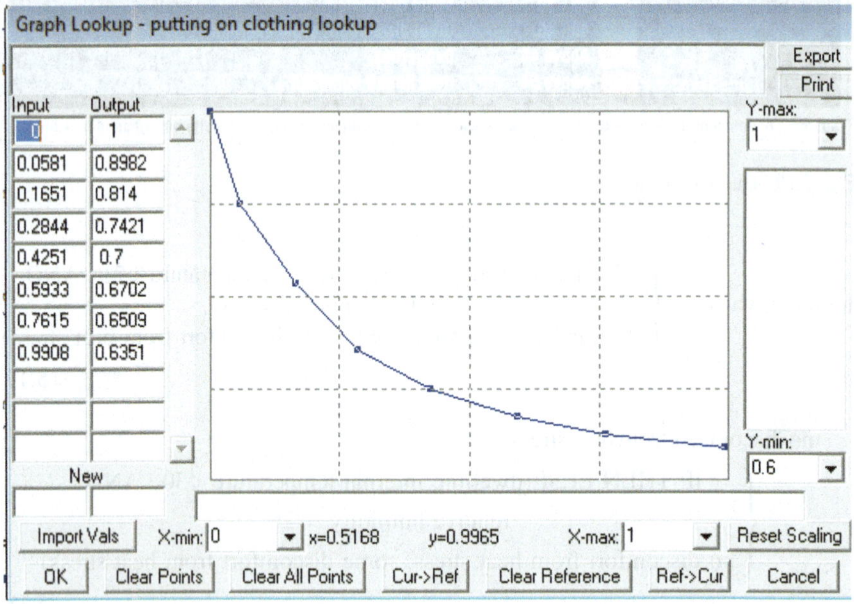

Fig. 5.21 Putting on clothing lookup

Table 5.7 Sample data for relative humidity

Variable	Unit of measurement	Minimum	Maximum	Mean	Standard error	Standard deviation
Relative humidity	Percentage	67	94	85.09	1.32	8.67

Met Office [10]

$$\text{some discomfort heat stress} = \begin{cases} \text{IF THEN ELSE(dwelling internal temperature} < 36 : \text{OR} : \\ \text{relative humidity} > 50, \\ \text{some discomfort, great discomfort)} \end{cases} \tag{5.19}$$

5.6.4 *SFD for Climatic-Economic-Energy Efficiency Interaction Module*

The structure of SFD for this module is not much different from its CLD as discussed in the preceding subsections, apart from establishing the relationships (in the form of equations) between different variables in the module. It is worthy of mention that no variable is represented as stock and flow. However, their impact is felt in other modules much like the household energy consumption module. To this end, this module shows the interactions of some energy efficiency, economic, and climatic variables that are included in the model. However, other variables of interest could be incorporated into the model by changing the structure of the model.

As shown in Fig. 5.22, the importance of household energy efficiency measures to household energy bills is highlighted. Also, the effects that unfavourable climatic conditions have on international energy prices and consequently on household energy bills are elaborated on. All these work together seamlessly as a system to give the effect of energy bills on energy consumption, which ultimately have effects on carbon emissions. An example of sample data driving this module is given in Table 5.8. Examples of major equations in this module are given in Eqs. (5.20)–(5.24).

$$\begin{aligned} &\text{effect of energy bills on energy consumption} = 1 - (1/((1 + (1/\text{average annual energy} \\ &\quad \text{bills}) * 100) + \text{effect of dwelling energy efficiency on energy bills}) - 0.9) \\ &\quad * (1 \quad -(1/(1 + \text{“\% increment on energy bills”}/\text{normal energy bills})))/(\text{effect of} \\ &\quad \text{unfavourable climatic effects on international energy prices}) \end{aligned} \tag{5.20}$$

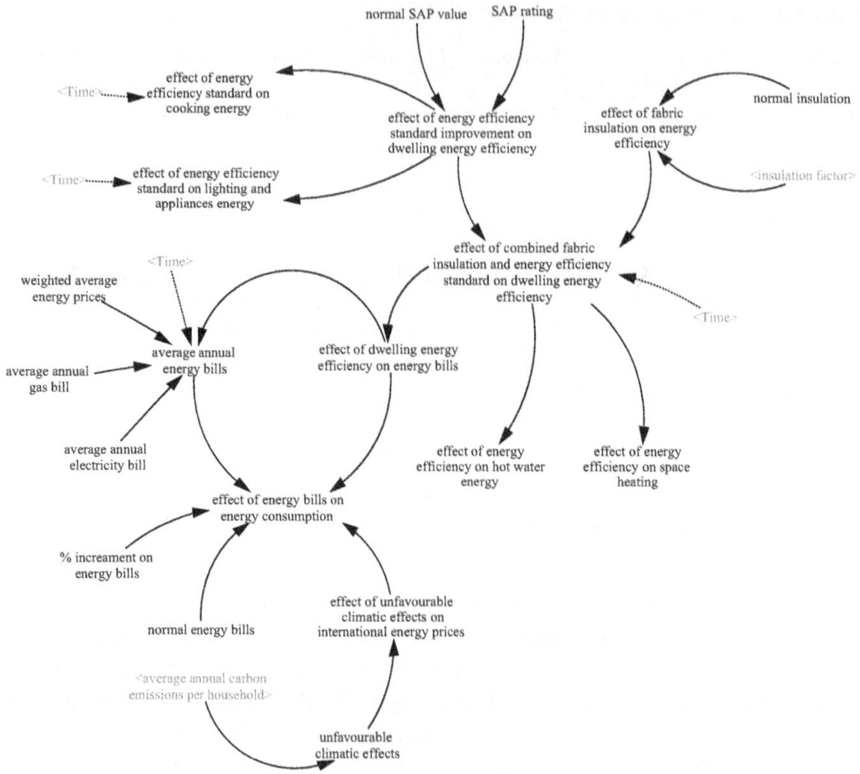

Fig. 5.22 SFD for economic-climatic-energy efficiency interaction module

Table 5.8 Sample data under climatic-economic-energy efficiency interaction module

Variable	Unit of measurement	Minimum	Maximum	Mean	Standard error	Standard deviation
Average annual gas bill	£	372	659	542.07	12.44	79.66
Average annual electricity bill	£	378	578	490.43	8.43	53.96
Weighted average energy prices	–	3.45	6.01	4.74	0.10	1.61
SAP rating	–	17.60	55.00	37.86	1.61	10.31

Palmer and Cooper [14]

$$\text{unfavourable climatic effects} = \text{SMOOTH}((1 - (1/\text{average annual carbon emissions per household})), 5) \tag{5.21}$$

$$\text{effect of dwelling energy efficiency on energy bills} = 1/\text{effect of combined fabric insulation and energy efficiency standard on dwelling energy efficiency} \tag{5.22}$$

$$\text{effect of energy efficiency standard improvement on dwelling energy efficiency} = \text{WITH LOOKUP(SAP rating/normal SAP value)} \tag{5.23}$$

$$\text{effect of fabric insulation on energy efficiency} = \text{WITH LOOKUP(insulation factor/normal insulation)} \tag{5.24}$$

5.6.5 SFD for Household Energy Consumption Module

In this module, household energy consumption is modelled. This is based on five different end-uses of energy (space heating energy consumption, hot water energy consumption, cooking energy consumption, lighting energy consumption, and appliances energy consumption). The details of SFD developed for 'space heating energy consumption', 'appliances energy consumption', 'hot water energy consumption', 'lighting energy consumption', and 'cooking energy consumption' are shown in Figs. 5.23, 5.24, 5.25, 5.26, and 5.27, respectively. As shown in Figs. 5.23, 5.24, 5.25, 5.26, and 5.27, it is necessary to state that the 'space heating carbon emissions', 'carbon emissions due to appliances energy', 'carbon emissions due to hot water usage', 'carbon emissions due to lighting', and 'carbon emissions due to cooking' are systematically modelled as accumulation of 'space heating energy consumption', 'appliances energy consumption', 'hot water energy consumption', 'lighting energy consumption', and 'cooking energy consumption', respectively, converted to carbon emissions through the use of 'energy to carbon conversion factor'. This method is used for all other household energy consumption end-uses. 'Average annual household energy consumption' (shown in Fig. 5.28) is therefore determined by adding all the household energy consumption stocks: space heating, hot water, cooking, lighting, and appliances energy consumption. Total annual household energy consumption is determined by multiplying the 'average annual energy consumption per household' by 'households'. An example of data driving the module is shown in Table 5.9. As example as well, equations relating to household energy consumption component of Fig. 5.23 are given (Eqs. 5.25–5.29). However, those relating to the carbon emission component are given in Sect. 5.6.6 (Eqs. 5.30–5.31).

(Palmer & Cooper, 2012)

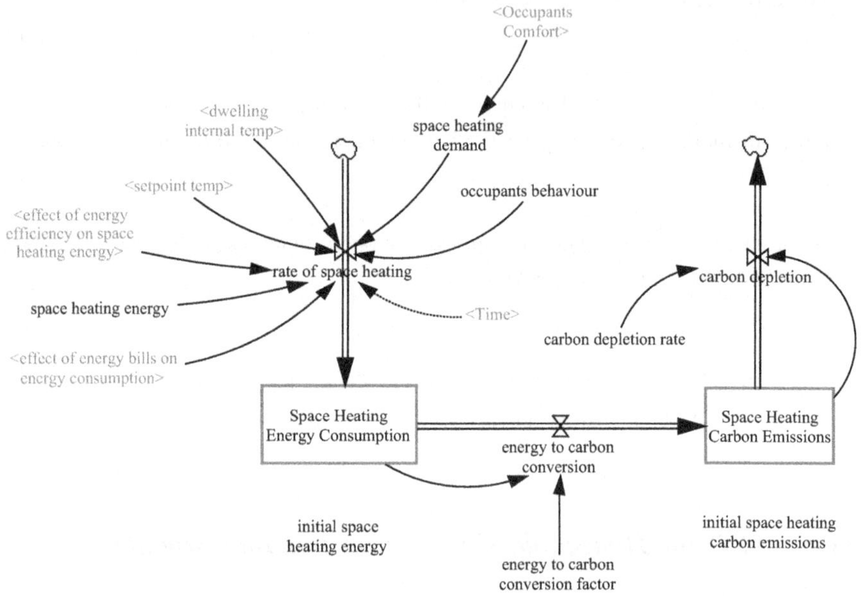

Fig. 5.23 SFD for space heating energy consumption and carbon emissions

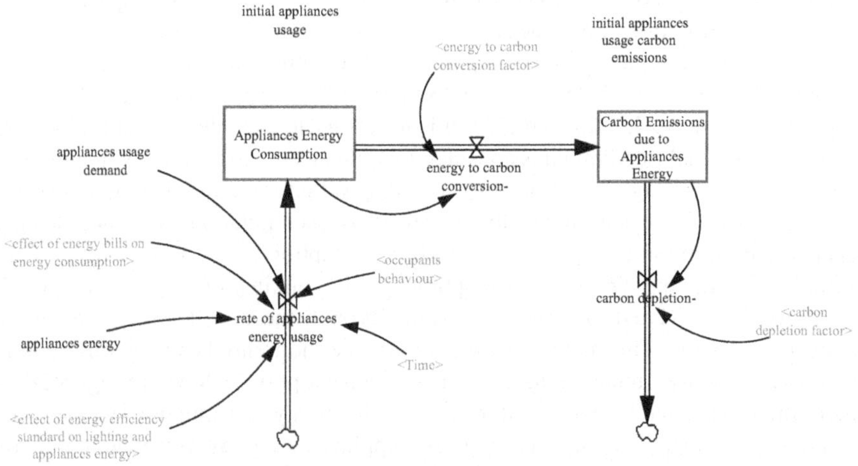

Fig. 5.24 SFD for appliances energy consumption and carbon emissions

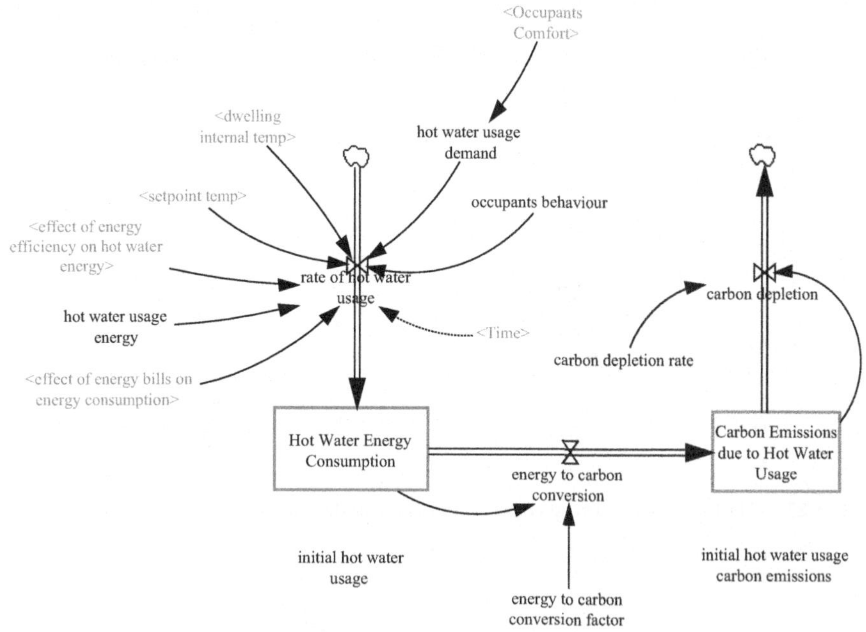

Fig. 5.25 SFD for hot water energy consumption and carbon emissions

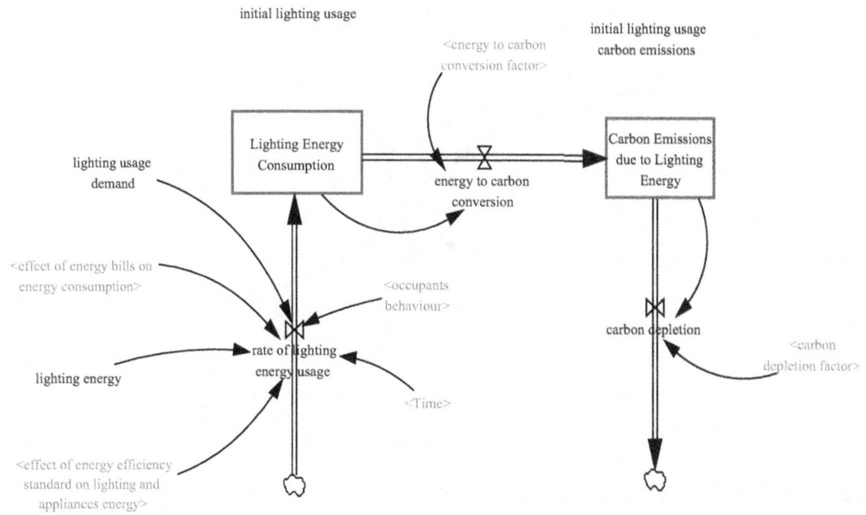

Fig. 5.26 SFD for lighting energy consumption and carbon emissions

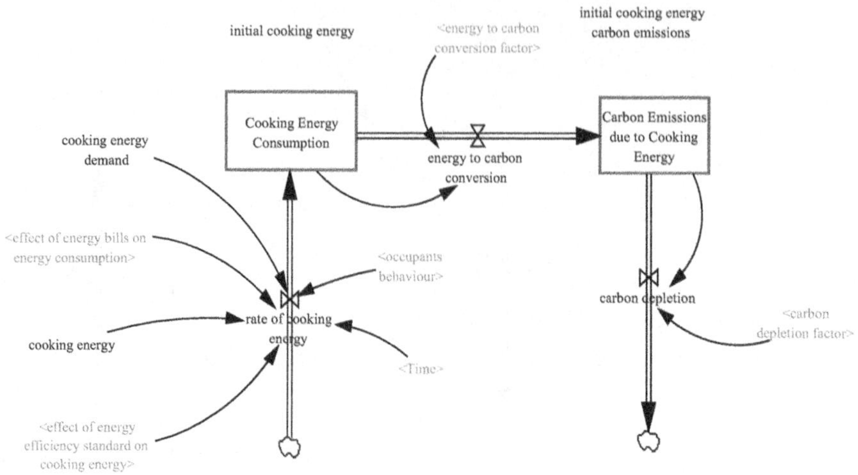

Fig. 5.27 SFD for cooking energy consumption and carbon emissions

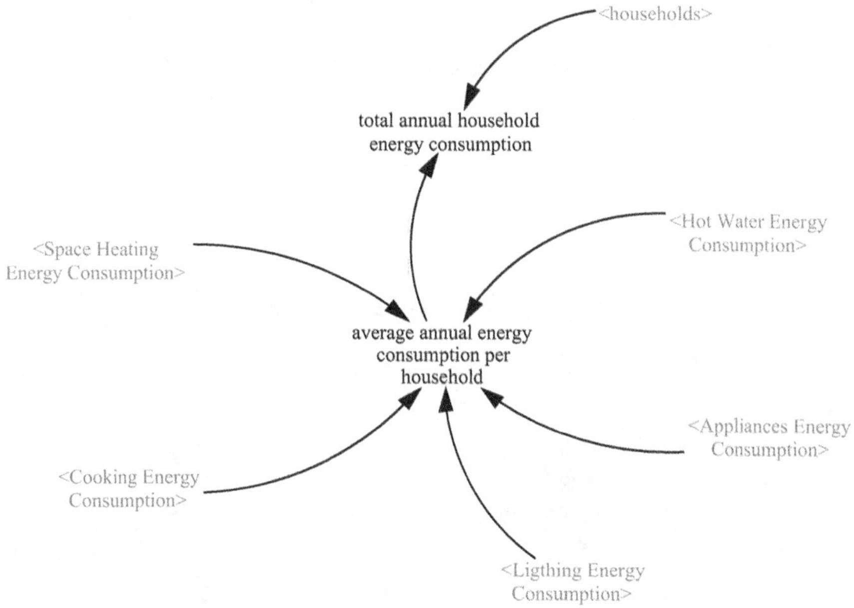

Fig. 5.28 SFD for household energy consumption

Table 5.9 Sample data for household energy by end-uses

Variable	Unit of measurement	Minimum	Maximum	Mean	Standard error	Standard deviation
Space heating	MWh	10.14	15.84	13.54	0.18	1.19
Hot water	MWh	3.03	6.64	4.78	0.17	1.10
Cooking	MWh	0.48	1.36	0.86	0.04	0.28
Lighting	MWh	0.55	0.69	0.65	0.01	0.04
Appliances	MWh	1.07	2.39	1.92	0.06	0.37

Palmer and Cooper [14]

$$\text{rate of space heating} = (\text{space heating energy} * \text{effect of energy efficiency on space heating/effect of energy bills on energy consumption} * 1.14 - 0.15 * \text{FORECAST}(\text{space heating energy} * 0.53, 39, 450)) * (0.60 * \text{setpoint temp})/\text{dwelling internal temp} \tag{5.25}$$

$$\text{Space Heating Energy Consumption}(t) = \text{INTEGRAL}[(\text{rate of space heating-energy to carbon conversion}), \text{initial space heating energy}(t_0)] \tag{5.26}$$

$$\text{energy to carbon conversion} = \text{Space Heating Energy Consumption} * \text{energy to carbon conversion factor} \tag{5.27}$$

$$\text{average annual energy consumption per household} = \text{Cooking Energy Consumption} + \text{Hot Water Energy Consumption} + \text{Lighting Energy Consumption} + \text{Space Heating Energy Consumption} + \text{Appliances Energy Consumption} \tag{5.28}$$

$$\text{total annual household energy consumption} = \text{average annual energy consumption per household} * \text{households}/10\char`^6 \tag{5.29}$$

5.6.6 SFD for Household Carbon Emission Module

As previously mentioned in Sect. 5.6.5 above, carbon emissions are modelled by converting energy consumption to carbon emissions through the use of a conversion factor termed 'energy to carbon conversion' (Figs. 5.23, 5.24, 5.25, 5.26, and

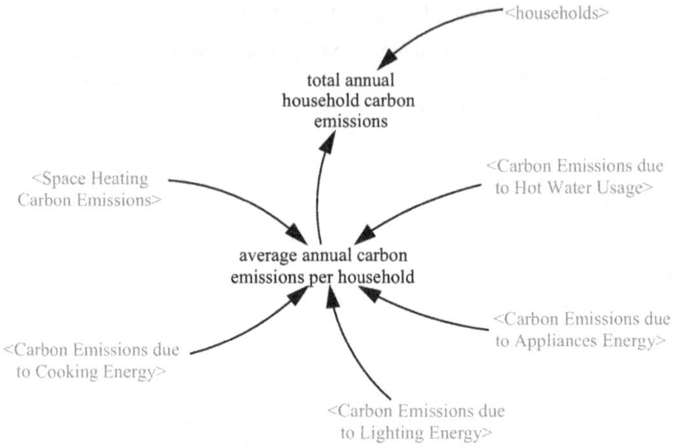

Fig. 5.29 SFD for household carbon emissions

5.27). It is important to state that the 'energy to carbon conversion factor' used in this model is assumed to be the conversion factor of energy to the carbon conversion factor of energy from electricity source. This is done for the sake of simplicity. Ideally, the energy conversion factor of different fuels (i.e. gas, oil, electricity) to meet household energy consumption by end-uses needs to be determined and applied appropriately. However, this is acknowledged as one of the limitations of this model. Average annual carbon emissions per household and total annual household carbon emissions (Fig. 5.29) are determined by the same approach as described under the household energy consumption module in Sect. 5.6.5.

$$\text{Space Heating Carbon Emissions}(t) = \text{INTEGRAL}[(\text{energy to carbon conversion-carbon depletion}), \text{initial space heating carbon emissions}(t_0)] \tag{5.30}$$

$$\text{carbon depletion} = \text{Space Heating Carbon Emissions} * \text{carbon depletion factor} \tag{5.31}$$

$$\begin{aligned} &\text{average annual carbon emissions per household} \\ &\quad = \text{Carbon Emissions due to Cooking Energy} + \text{Carbon Emissions due to Hot Water Usage} \\ &\quad + \text{Carbon Emissions due to Lighting Energy} + \text{Carbon Emissions due to Appliances Energy} \\ &\quad + \text{Space Heating Carbon Emissions} \end{aligned} \tag{5.32}$$

total annual household carbon emissions = average annual carbon emissions per household * households/10^6 (5.33)

5.7 Details of Variables Excluded from the Model

In Sect. 5.3 of this book, the boundary for the model was discussed. It was highlighted that there is a need to have a model boundary chart that detailed the variables included in the model in the form of endogenous and exogenous variables and those that are excluded. Considering the type of complexity involved in the system being modelled in this book, some variables relating to occupants' behaviour such as 'occupants' social class influence' and 'occupants' cultural influence' among others (Table 5.2) are excluded from the model. This is mainly because of the fact that 'occupants' behaviour' in the developed model is currently modelled exogenously based on the assumption that 'occupants' behaviour' externally affects household energy consumption. Inclusion of these variables will mean that a great deal of time will be committed to conducting social research relating to different influences on 'occupants' behaviour' leading to modelling the 'occupants' behaviour' endogenously. This is, however, seen as a limitation of the research reported in this book.

Additionally, some variables relating to the physical characteristics of dwellings such as dwelling exposure, and air changes, amongst others are also excluded from the model. These variables are specific to individual dwellings. Also, it is necessary to strike a balance between aggregated and disaggregated variables to be included in the model because of the target audience of the model output, which in this case are the energy policy decision makers. Furthermore, some variables relating to external environment such as political uncertainties and energy securities are not modelled considering the scope of the research and their noninclusion signifies the potential of the model to explore an array of issues.

5.8 Summary

This chapter has extensively described and discussed the model conceptualisation and formulation, including the model boundary based on the variables incorporated and excluded from the model. The chapter discussed in detail information about those who participated in the model conceptualisation process. The mental model developed by these individuals was captured in the form of knowledge elicitation in order to improve and validate the causal diagrams drawn by the modeller. The final causal diagrams developed for the model were the ones represented and described in this book. The chapter also described and discussed the model formulation in the form of stock and flow diagrams for all the modules. The key algorithms relating the variables were also given. The next chapter of this book will discuss the behaviour of the simulation run based on the 'baseline' scenario.

References

1. American Society of Heating, Refrigerating and Air Conditioning Engineers (2004) Thermal environmental conditions for human occupancy. ASHRAE Standard 55-2004, Atlanta, USA
2. Building Research Establishment (2011) The government's standard assessment procedure for energy rating of dwellings. 2009 edition incorporating RdSAP 2009, Watford, UK
3. Building Research Establishment (2012) The government's standard assessment procedure for energy rating of dwellings, Garston, Watford
4. Chartered Institution of Building Services Engineers (2006a) Comfort. CIBSE Knowledge Series KS6, London
5. Chartered Institution of Building Services Engineers (2006b) Environmental design. CIBSE Guide A, London
6. Coyle RG (1997) System dynamics modelling: a practical approach. Chapman & Hall, London
7. Fanger OL (1970) Thermal comfort: Analysis and applications in environmental engineering, McGraw-Hill
8. Forrester JW (1971) World Dynamics. Cambridge, MA: Wright-Allen Press, Inc
9. International Standard Organisation (1994) Moderate thermal environments—determinants of the PMV and PPD indices and specification of the conditions for thermal comfort. ISO 7730, Geneva
10. Met Office (2013) Climate data. Available at www.metoffice.gov.uk (Accessed on September 18, 2013)
11. Mohammed RE (2007) An exploratory system dynamics model to investigate the relationships between errors that occur in construction documents in Saudi Arabia and their possible causes. PhD thesis, Heriot-Watt University, Edinburgh, UK
12. Motawa I, Oladokun M (2015) A model for the complexity of household energy. Energy Build 87:313–323
13. Oladokun MG (2014) Dynamic modelling of the sociotechnical systems of household energy consumption and carbon emissions. Ph.D. thesis, Heriot-Watt University, Edinburgh, UK
14. Palmer J, Cooper I (2012) United Kingdom housing energy fact file. DECC, London
15. Pruyt E (2013) Small system dynamics models for big issues: triple jump towards real-world dynamic complexity. TU Delft Library, Delft, The Netherlands
16. Sterman JD (2000) Business dynamics: systems thinking and modelling for a complex world. McGraw-Hill, New York

Chapter 6
Simulation of Energy and Carbon Emissions in Housing

6.1 Introduction

This chapter presents the model behaviour based on the 'baseline' scenario. It communicates the most likely way in which the household energy and carbon emissions of the UK housing stock will evolve over the years starting from 1970 until 2050. The 'baseline' scenario assumes the continuation of the trends depicted by historical data based on the current trends of energy efficiency measures, 'standard' consumption behaviour, and energy prices. The chapter first describes the general assumption underpinning the 'baseline' scenario. This is followed by a discussion of the insights from the model in terms of the behaviour generated based on different modules represented in the model. The model testing and validation concludes the chapter.

6.2 General Assumptions About the Simulation

In conducting simulation of household energy and carbon emissions, this book made use of the 'baseline' scenario to serve as the reference case to all other scenarios formulated in this book as will be discussed in Chap. 7. It is the reference point to all other scenarios and forms the basis for comparison. In this scenario, it is assumed that there will be few or no changes to the existing policy in energy efficiency with an assumption that the existing policy will continue to be in operation without any further policy intervention to salvage the situation. This by implication means that the pattern of energy efficiency measures for dwellings, heating systems, cooking, lighting, and appliances based on available historical data will continue to follow the current trend, without any specific efforts to upturn the trend.

M. G. Oladokun and C. O. Aigbavboa, *Simulation-Based Analysis of Energy and Carbon Emissions in the Housing Sector*, Green Energy and Technology,
https://doi.org/10.1007/978-3-319-75346-1_6

Also, the 'baseline' scenario assumed three behavioural classifications for the model in this book. These are 'frugal', 'profligate', and 'standard' behavioural classifications. 'Standard' here means that the consumption behaviour of occupants is assumed to be a mid-way between the profligate and frugal consumption behaviours. The 'baseline' scenario therefore assumes that the energy consumption behaviour of occupants of dwellings in the UK is 'standard'. As such, the internal temperature of dwellings is assumed to have a set-point of 19 °C. Further, the scenario assumes that any change to energy bills will not significantly affect the energy consumption behaviour of the occupants as the 'standard' consumption behaviour will be maintained. Both the number of households and average household size in the UK impact on the energy consumption profile of the UK housing stock, and as such their future trends as emanate from the output of UK population are maintained for the scenario.

6.3 Behaviour Analysis of Precursor Modules to the Household Energy Consumption and Carbon Emission Modules

It is important to present and discuss the simulation results of the modules preceding the household energy consumption and carbon emissions modules. This is to have an understanding of the insights generated by key variables in these modules and see how those insights will influence the behaviour analysis of household energy consumption and carbon emissions modules. As such, the behaviour analyses of key variables in the population/household, dwelling internal heat, occupants' thermal comfort, and climatic-economic-energy efficiency interaction modules are discussed in the following subsections.

6.3.1 Behaviour Analysis of Some Variables in Population/Household Module

It has been shown in the preceding chapter that one of the modules that constitute the model in this book is the population/household module. The importance of this module cannot be overemphasised as the number of households plays a major role in accurately estimating the amount of energy consumption in the entire UK's housing stock. This is mainly due to the fact that energy consumption in homes is driven by the quest for energy services such as comfort by occupants. This means that actual energy required in meeting these services reflects, for example, the type of services required by the occupants and the factors relating to fabric insulation in homes, heating systems, and appliances use, among others. Invariably, occupants consume energy as a result of their seeking comfort at home. Notwithstanding this, it is not surprising that household energy consumption is strongly influenced by the population, the number of households, and the average household size. For example, the amount of household energy consumption attributable to hot water consumption and usage of some appliances is greatly influenced by the household size. However, there is a minimum level of household energy consumption applicable to each household as the operation of some energy-consuming appliances such as fridges, fridge-freezers, or freezers, do not depend on the household size.

The model outputs in this module are presented in Figs. 6.1, 6.2, and 6.3 to, respectively, illustrate the behaviours of total UK population, total number of UK households, and average household size in the UK ranging from 1970 to 2050. The model shows that the total UK population is on an upward trend till 2050 (Fig. 6.1). The model indicates that the UK population of 55.63 million in 1970 will grow to about 69.78 million by the year 2050. This figure shows a yearly increase of 0.31% on average. Comparing this figure with data from the Office of National Statistics (2013) suggests that the UK population recorded an average annual growth of 0.28% between the year 1970 and 2010. Within the same time horizon of 1970 and 2010, the model output shows an average yearly population growth of 0.26%. The slight difference in the two estimates can be attributed to the methods used in the development of the models.

Additionally, Fig. 6.2 shows that the average number of households grows steeply on a yearly basis when compared to the steepness of the population output as shown in Fig. 6.1. The result of the model indicates that the number of households of 18.78 million will grow to about 34.10 million by the year 2050. This reflects an overall average yearly increment of about 1.02%. The growth in the number of households is an indication of the rising number of smaller households, which reflects that more people tend to live all alone or in smaller family sizes. This does not only have implications for the adequate provision of housing for the

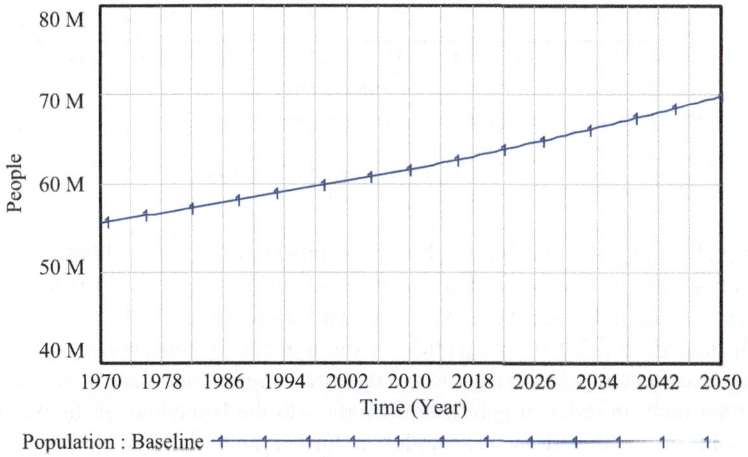

Fig. 6.1 Projected total UK population under the 'baseline' scenario

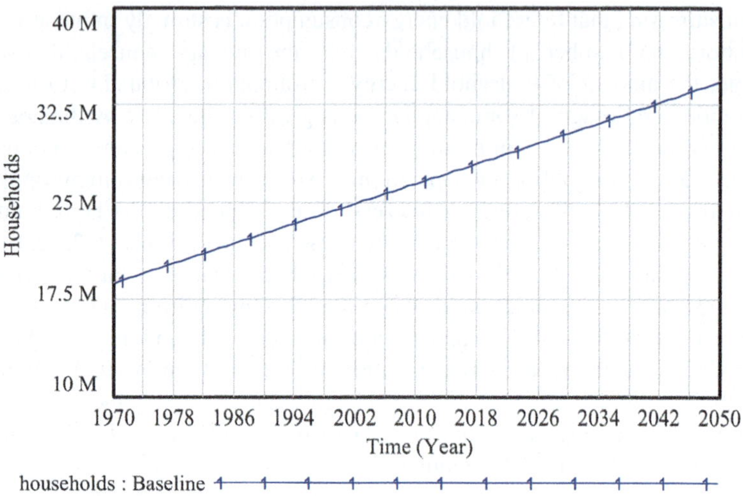

Fig. 6.2 Projected total number of UK households under the 'baseline' scenario

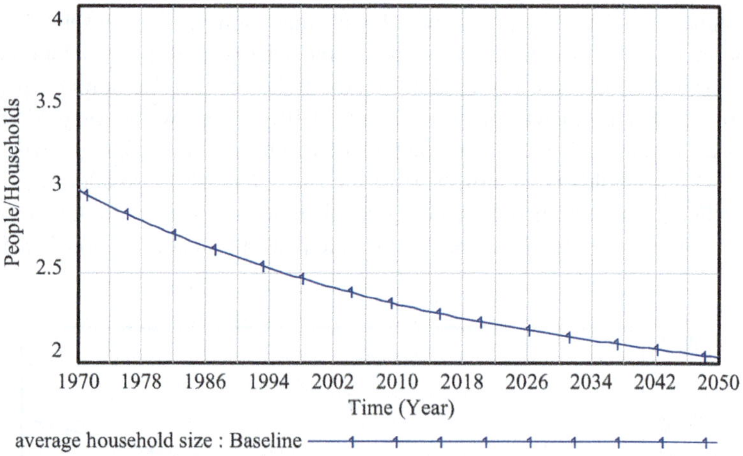

Fig. 6.3 Average UK household size under the 'baseline' scenario

citizenry, but alters the housing stock energy consumption profile of homes. This by implication means that per capita energy consumption will tend to grow as the number of households increases with attendant decrease in the average household size. It then shows that the projection of the number of households is the key to accurately estimating the household energy consumption and carbon emissions.

Furthermore, in order to gain more insight into the behaviour of the number of households as produced by the model, a further analysis was conducted at an interval of ten years from 1970. The results of this study are compared with those of

Johnston [11] and Palmer and Cooper [14] as shown in Table 6.1. The behaviour of this study indicates an increase in the number of households on a yearly basis, but with a decline in the level of this growth until 2050. The result shows almost the same downward trend as the output of the Johnston [11] model. The major difference between the two models lies in the region of 2040–2050, where Johnston [11] specifically stated the trend of this region is fixed (and not the result of analysis) based on the assumption that the number of households will not change in those years. This assumption then explains the difference in the two models. Correspondingly, when the output of this model is compared with that of Palmer and Cooper [14], which is based on the ONS available data till 2010, the results follow a 'lumpy' trend with a combination of peaks and troughs, but overall the number of households grow at 0.99% yearly, whereas the results of this model show an annual growth of 1.02% (Table 6.1).

Equally, it is expedient to explain some insights shown by the behaviour of average household size trend as depicted in Fig. 6.3. The model result suggests that the trend of the UK household size on average follows a downward trajectory with the slope of the graph between 1970 and 2020 a little steeper than that of between 2020 and 2050. That is, the average household size declines steadily from 2.96 in 1970 to 2.04 in 2050. The reason for this trend may be attributed to the growth in the number of single-person households without children as previously advanced. Traditionally, households, for example, used to comprise married couples living together with their dependent children, but a decline in this kind of proportion may also likely be responsible for the simulation behaviour. Furthermore, it may be that there is a growth in the number of households consisting of married couples without dependent children or an increase in the proportion of single-parent households. All these factors may be responsible for the decline witnessed in average household size in the model. The results of this model are consistent with historical data as

Table 6.1 Average yearly percentage increase/decrease in the number of households

Year	Johnston [11] (%)	Palmer and Cooper [14] (%)	This study (%)
1970–1980	–	0.82	0.93
1980–1990	–	0.88	0.84
1990–2000	0.76[a]	0.72	0.77
2000–2010	0.70	0.81	0.71
2010–2020	0.68	–	0.67
2020–2030	0.53	–	0.63
2030–2040	0.37	–	0.59
2040–2050	−0.09	–	0.55
Overall yearly average	0.56[a,b]	0.99[b]	1.02[b]

[a]The year starts from 1996; [b]was computed based on [(final year value − base year value)/number of years * 100%]

contained in Palmer and Cooper [14] and the output of Johnston's [11] model. The next section discusses the behaviour of key variables within the dwelling internal heat module.

6.3.2 *Behaviour Analysis of Some Variables Within the Dwelling Internal Heat Module*

In this module, the only variable that changes from one scenario to another is the set-point temperature designated 'set-point temp' within the model. The baseline value assumed for this variable is 19 °C. This part of the book therefore presents the behaviour of key variables within the dwelling internal heat module. As mentioned in Chap. 5 of this book, dwelling internal heat is mainly required to model its impact on occupants' comfort and consequently on space heating and hot water requirements of the occupants. The dwelling internal heat is principally influenced by the amount of heat introduced into the dwelling and determines the dwelling internal temperature. The total dwelling heat gains (DHGs) for the entire UK housing stock are modelled from six different sources as explained in Chap. 5 to include DHGs due to appliances, artificial lighting, cooking, number of people (metabolic heat gains), water heating, and solar gains based on BRE [4]. The degree of infiltration into/out of the dwelling is modelled and captured as heat losses.

The insights generated from the simulation for DHGs due to appliances, artificial lighting, cooking, water heating, heat losses, and the number of people are shown in Fig. 6.4. It should be noted that the graph in Fig. 6.4 applies a multiscale method for compact display of the results. In the multiscale graph, the digits '−80' and '−200', respectively, represent the upper and lower values of 'heat losses'. Similarly, the values of '400' and '200', respectively, denote the upper and lower values of the second variable designated '2', which in this case is 'DHG due to appliances less cooking'; among others. The insights generated from the simulation run for the DHGs due to appliances, artificial lighting, cooking, number of people, and water heating indicate that all of them reveal the same trends of gently declining. However, these patterns are not too distant from the behaviour displayed by the average household size as shown previously in Fig. 6.3. The main reason that could be advanced for these insights is the dominant effect of the average household size of the UK housing stock that has been on a declining trend from an average of 2.96 in 1970 to 2.04 by 2050 as the model results suggest.

Furthermore, the results as shown in Fig. 6.4 give an insight into the pattern of behaviour expected from the heat losses from the UK housing stock. The result indicates a decrease in the value of heat losses from dwellings. The main explanation for this insight could be attributed to the fact that different schemes are in place being trialled to improve the airtightness in dwellings, according to the melioration of fabric insulation of dwellings [12]. Correspondingly, another important component of DHGs that is not captured in Fig. 6.4 are solar gains due to

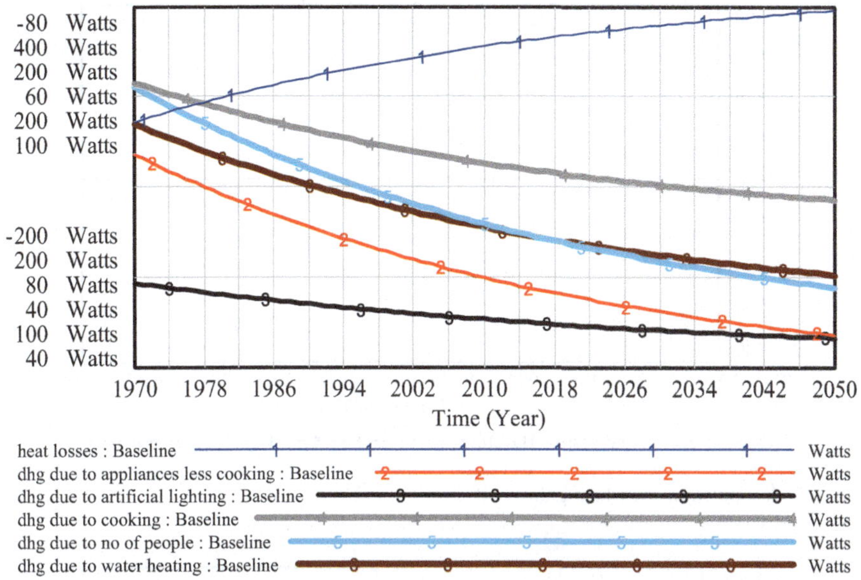

Fig. 6.4 Insights from some of the key variables of dwelling internal heat module

the limitation of the software that cannot combine more than six variables in a graph. However, this is shown in Fig. 6.5. Again, this follows the method described in SAP 2012 [4] as explained in Sect. 5.6.2 of Chap. 5. Using this method involves making a number of assumptions as presented in BRE [4] regarding solar flux, solar transmittance factor for glazing, frame factor, average solar access factor, and average area of openings in dwellings. All these parameters are constant, and this is why the trend exhibited by solar gains is a perfect horizontal trend as shown in Fig. 6.5. The total DHGs as a result of the summation of these different gains produce gentle declining behaviour as illustrated in Fig. 6.6.

Based on the insights above, the total DHGs accumulate the artificial heat transfer, which was modelled as a flow as shown in Fig. 5.17 of Chap. 5. This is one of different variables that are driving the dwelling internal heat that is simulated as an accumulation of natural and artificial heat transfers in and outside the dwellings. This then provides the thermodynamics of heat balance within the dwelling. Following the same way, the average dwelling internal temperature is modelled as well. The simulation results show that the dwelling internal heat tends to increase over time (Fig. 6.7) and likewise the average dwelling internal temperature (Fig. 6.8). The simulation results suggest that the trends witnessed from 1970 would continue until the year 2050. Another insight from the simulation results of Figs. 6.7 and 6.8 is that the trends follow the same pattern of growth with 'troughs' and 'peaks', respectively, representing the periods of extreme and milder winter weather in the UK [12]. The extreme weather conditions of the year 2010 may serve as an example of a drop or 'trough' in temperature, which

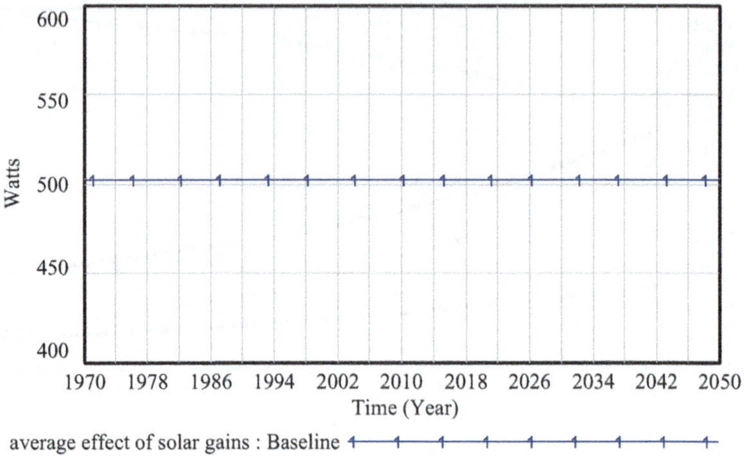

Fig. 6.5 Insight from solar heat gains

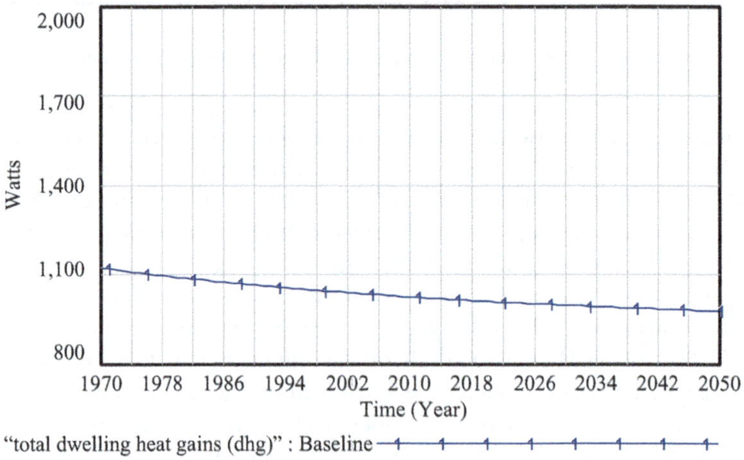

Fig. 6.6 Total dwelling heat gains under the 'baseline' scenario

correspondingly affects 'dwelling internal heat' and 'dwelling internal temperature' as explained above. The model then suggests that the weather condition will improve over time. Therefore, this is reflected in picking up again as the behaviours of these two variables suggest. These results are consistent with the output of a number of studies [1, 5] in the UK suggesting the risk of overheating in UK homes in years to come as well as the threat of the global climate warming effects. Furthermore, the simulation outputs suggest some interesting insights in the area of airtightness of dwellings in the form of infiltration. The simulation results show that this will reduce over time, possibly as a result of improved fabric insulation, leading

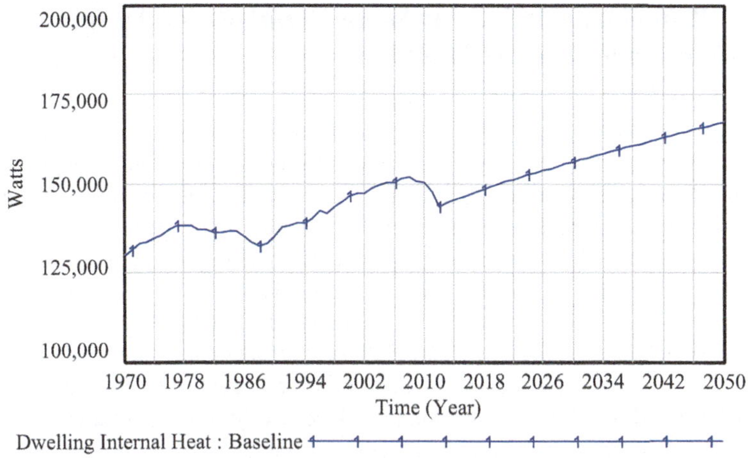

Fig. 6.7 Dwelling internal heat under the 'baseline' scenario

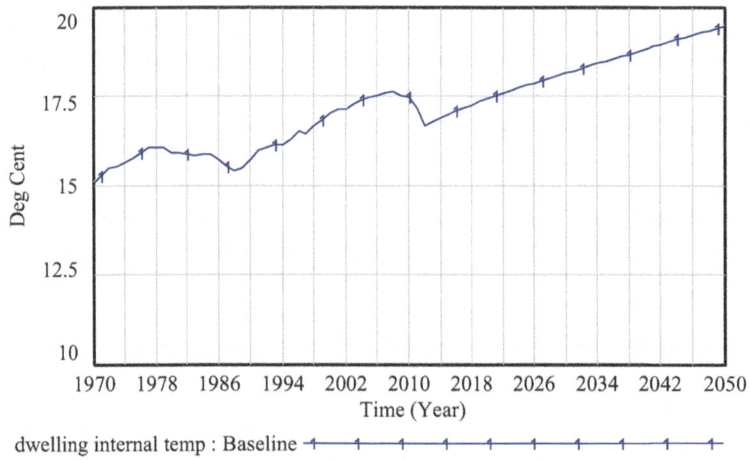

Fig. 6.8 Dwelling average internal temperature under the 'baseline' scenario

to reduced wind forces, which create pressure differences within dwellings. This by implication will see dwellings in the UK retaining more internal heat due to an increase in space heating and internal temperature.

For example, the model results envisaged that the average dwelling internal temperature will continue to increase based on the desire of occupants to improve thermal comfort by raising the temperature set-point as long as energy prices are kept low. It is necessary to emphasise that the increment will not continue indefinitely. As such, a saturation level will be reached in such a way that any further rise in internal temperature would be a discomfort to occupants.

6.3.3 Behaviour Analysis of Some Variables in Occupants' Thermal Comfort Module

The outputs expected from some key variables of the occupants' thermal comfort module based on the simulation run are presented here. The average relative humidity as obtained from the UK Met Office as well as the result of the average dwelling internal temperature from Sect. 6.3.2 is the main input into this module. Once the simulation has been run, the perceived dwelling temperature is modelled and consequently the occupants' comfort. As explained in Sect. 5.6.3 of Chap. 5, the perceived dwelling temperature is modelled qualitatively based on the humidex chart referred to in Fig. 5.19 of Chap. 5. The chart takes the values of the average dwelling internal temperature and average relative humidity to estimate the humidex value, which in turn produces the perceived dwelling temperature as shown in Fig. 6.9. This is in combination with other variables such as the probability of window opening, probability of putting on clothing, occupants' metabolic build-up, and occupants' comfort as discussed in Sect. 5.6.3. From Fig. 6.9, it can be seen that the trend noticed in the output of the perceived dwelling temperature variable follows that of the average dwelling internal temperature. This indicates that there is a gentle growth in the perceived dwelling temperature. By implication, it means that once the perceived dwelling temperature by the occupants rises, occupants will definitely react to this. It may either be to open window(s) or put on more clothing in order to reach the required thermal comfort.

It is explained in Sect. 5.6.3 of Chap. 5 that the probabilities of putting on clothing and opening windows are qualitatively modelled from the 'putting on clothing lookup' and 'window opening lookup'. Therefore, the insights into the chances of occupants' putting on more clothing or opening window(s) that shaped

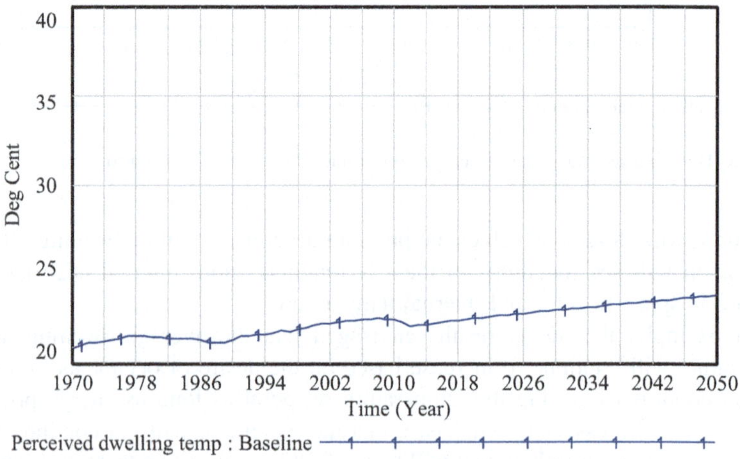

Fig. 6.9 Perceived dwelling temperature under the 'baseline' scenario

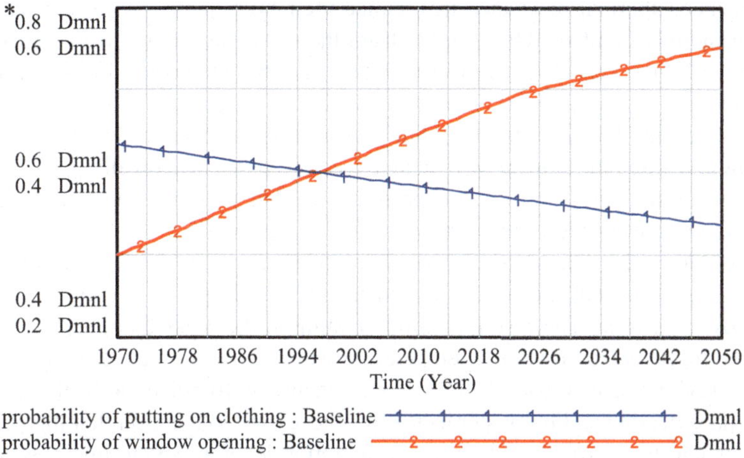

Fig. 6.10 Probabilities of putting on clothing and window opening under the 'baseline' scenario. Asterisk Dmnl—dimensionless

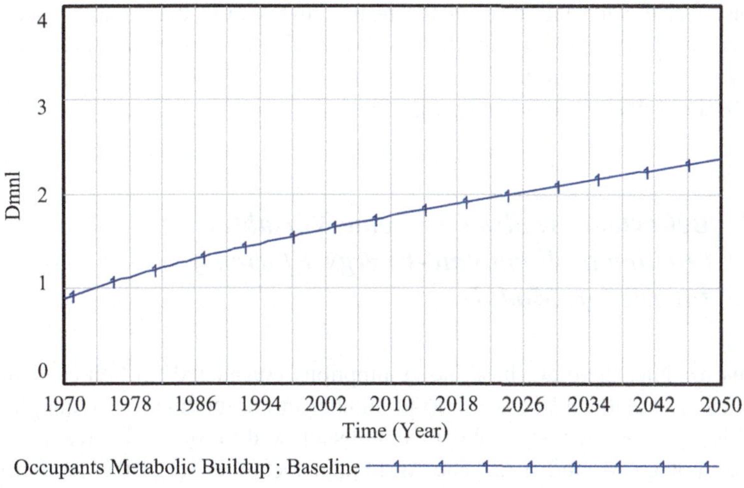

Fig. 6.11 Occupants' metabolic build-up under the 'baseline' scenario

the pattern of behaviour of perceived dwelling temperature (see Fig. 6.9), occupants' metabolic build-up (see Fig. 6.11), and occupants' comfort (see Fig. 6.12) as generated from the simulation are shown in Fig. 6.10. Since the result of the perceived dwelling temperature shows a gentle slope in growth (see Fig. 6.10), this implies that occupants will tend to open their dwellings window(s) in order to reach the required thermal comfort. Also, it is possible for them to remove dense clothing, which obviously has high thermal resistance, and purposely put on light clothing with reduced thermal resistance to regulate their body temperature and reach the desired thermal comfort. The insights from the model as shown in Fig. 6.10 indicate that over the years, starting from 1970 until 2050, the probability of putting on clothing with increased thermal resistance will tend to decline on the average, while at the same time horizon, the probability of occupants' opening windows to reach the required thermal comfort will increase as the perceived dwelling temperature increases. Again, these results are profound in that they are consistent with the global climate warming predictions.

Considering the discussion above, the simulation results depicted in Figs. 6.11 and 6.12 indicate that the occupants' metabolic build-up and occupants' comfort will grow over time. The results are illuminating in the sense that they are tailored towards the perceived dwelling temperature which is on the increase as suggested by the previous results discussed. This may then lead to a decline in the quest for more space heating and hot water usage. It is necessary to mention that there will be a time when these growths will reach a saturation level at which time they tend to decline. As can be seen from the simulation results, the outputs produce no such plausible insight, possibly due to the fact that occupants' comfort is being regulated by the two aforementioned actions of the occupants—window opening and putting on of clothing. Another profound insight from this simulation may be the introduction of artificial ventilation in the near future should the value of occupants' comfort grow prodigiously to the level that the two aforesaid actions of the occupants, as assumed in this model, are no longer valid. This may, however, add to the household energy consumption profile.

6.3.4 Behaviour Analysis of Some Variables in Climatic-Economic-Energy Efficiency Interaction Module

For this module, there are baseline assumptions considered for some of the variables that change from one scenario to another in the module. For example, 'SAP rating' is based on the historical data and this allowed to follow the historical trend, but does not go beyond 0.75. The 'insulation factor' is set at 0% as there is no change to the historical trend. Also, the value of the '% increment on energy bills' is set at 0% as no increment is assumed for the 'baseline' scenario. This section of the book, therefore, gives the results of 'baseline' scenario analysis performed for the climate-economic-energy efficiency interaction module. This is to show the effects of key variables in the module on energy consumption and carbon emissions as will be discussed later in this chapter of the book. Figure 6.13 shows the profile of unfavourable climatic effects, which is internally generated based on the interaction of different variables in the model. Within the energy-related research community, it is widely accepted that changes in the global climate are as a result of an increase in carbon emissions in the atmosphere [17]. An increase in carbon emissions is then

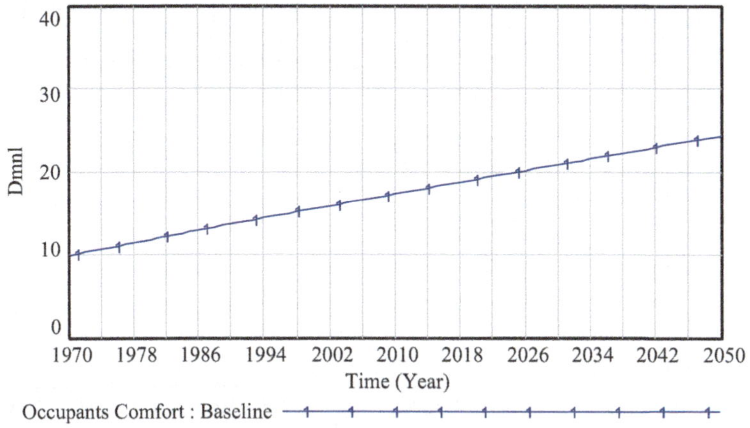

Fig. 6.12 Occupants' comfort under the 'baseline' scenario

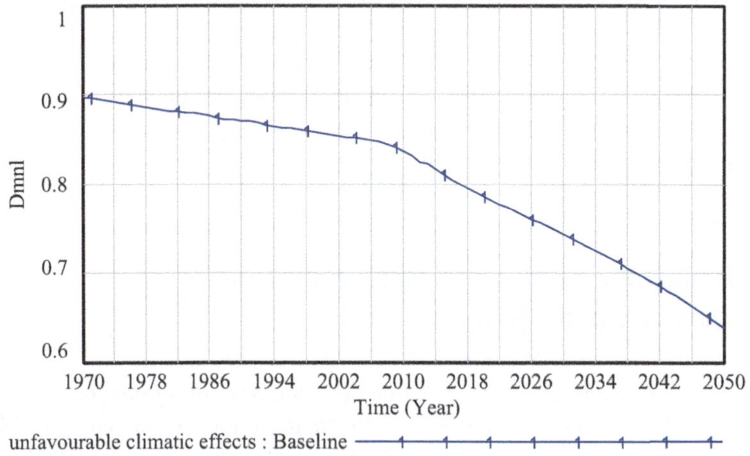

Fig. 6.13 Behaviour of unfavourable climatic effects

likely to have unfavourable climatic effects. The results of this model shown in Fig. 6.13 suggest that unfavourable climatic effects tend to decline. The reason behind this insight is that carbon emissions tend to be reduced in atmosphere. This may be due to different efforts geared towards reducing carbon emissions released into the atmosphere.

The effect of energy efficiency standard improvement on dwelling energy efficiency was qualitatively modelled to show how the model will respond to changes in the energy efficiency standard as dictated by the values of the SAP rating of dwellings. As shown in Fig. 6.14, the model result suggests that the effect of energy

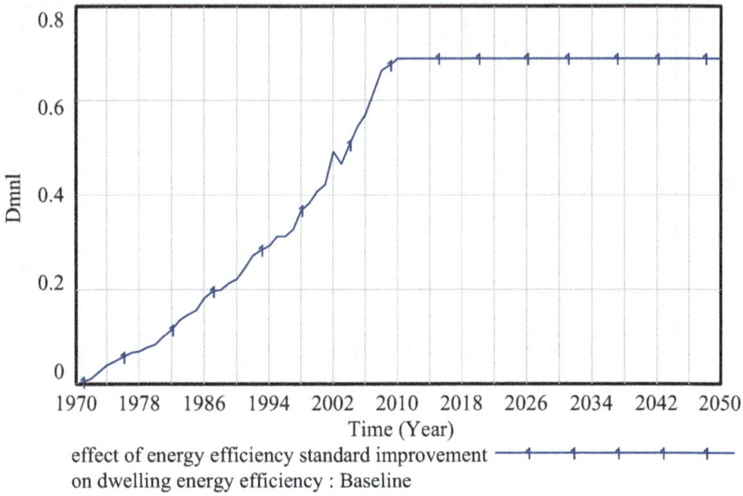

Fig. 6.14 Behaviour of effect of energy efficiency standard improvement on dwelling energy efficiency

efficiency standard improvement on dwelling energy efficiency will tend to improve, but will not rise above a certain value (0.75) as the model assumption for 'SAP rating' suggests.

6.4 Behaviour Analysis of Household Energy Consumption Module

This section of the book gives the behaviour analysis of simulation run for the 'baseline' scenario. For this module, 'occupants behaviour' is the only variable that can serve as the policy lever and hence changes for all other scenarios. As discussed in Sect. 6.2 of this book, the 'occupants' behaviour' variable can assume three possible values of '1' to '3' with '1' being frugal consumption, '2' standards consumption, and '3' profligate consumption. The simulation run here assumed a standard consumption pattern. The insights into energy consumption attributable to the 'baseline' scenario are therefore presented and discussed in this section. These insights are discussed based on the end-use household energy consumption. To this end, the following subsections show and discuss the trend of energy use for space heating, hot water, cooking, lighting, appliances as well as the total household energy consumption for the entire UK housing stock.

6.4.1 *Behaviour Analysis of Space Heating Energy Consumption*

The behaviour over time of the average space heating energy consumption per household in the UK is shown in Fig. 6.15. The figure shows that the household energy consumption attributable to space heating energy is high and may likely be the biggest portion of the total UK household energy. This is because its average annual value has been hovering around 15 MWh per household for the first four decades. Within this period, the space heating energy was in an upward trajectory until the year 2004 when it began to decline, with an exception of the year 2010 (which is due to the bad weather conditions of 2010). The reason that could be advanced for the growth in energy over the first four decades may be the behavioural attitude of occupants as they seek more thermal comfort at home, thereby raising the internal temperature of their homes. It may also be due to homes' extensions over the years that result in increased heated volume, which significantly adds to the space heating energy.

Based on the assumptions for the 'baseline' scenario, the model forecasts that the space heating energy will continue to follow a downward trend from the year 2004 until 2050 owing to improvements in energy efficiency (SAP rating) as a result of stringent building regulations and other areas of government campaigns, including occupants' behavioural change towards energy consumption. Further to these reasons, the downward simulation results may be due to a rise in energy costs since 2004 as advanced by Summerfield et al. [19] and may as well be unconnected to milder winters [14].

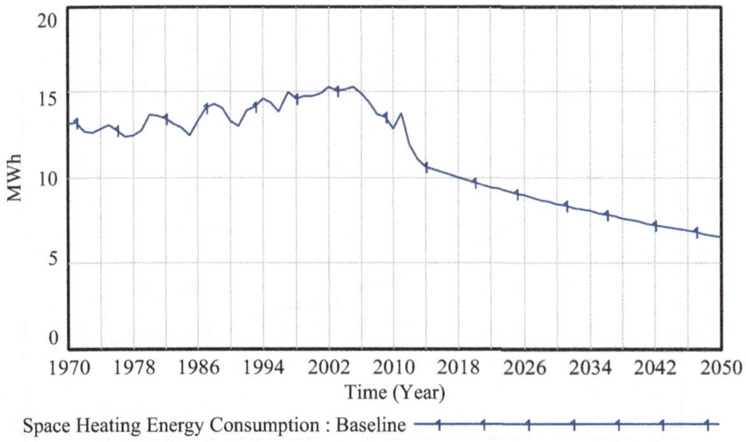

Fig. 6.15 Average space heating energy consumption per household

6.4.2 Behaviour Analysis of Hot Water Energy Consumption

The simulation result suggests that the hot water energy use for the average UK homes has been significantly reduced and followed a downward trajectory since 1970 as depicted in Fig. 6.16. There are some reasons that can be advanced for this pattern. One of these reasons could be the reduction in heat loss from hot water tanks (in terms of improvement in energy efficiency 'SAP rating') according to Oladokun and Odesola [13]. This improvement may be due to the ameliorated lagging of hot water pipes and tanks together with the improvements being witnessed in the household heating systems as a result of building regulations changes. A further probe into the behaviour of the model indicates that the slope of the trend changed slightly around 2014 and will follow this new trend until the year 2050. Should the trend follow the slope of the graph since 1970 until 2014 as shown in Fig. 6.16, it may mean that by 2040, the average energy consumption for hot water would reach net zero, which is practically impossible. Irrespective of the demand for cuts in household energy consumption, it will not translate to mean that no hot water would be required at homes in years to come as there will always be a minimum amount of hot water energy required for each household.

6.4.3 Behaviour Analysis of Cooking Energy Consumption

The insight as revealed by the simulation output for the average cooking energy consumption per household is illustrated in Fig. 6.17. Generally, the pattern

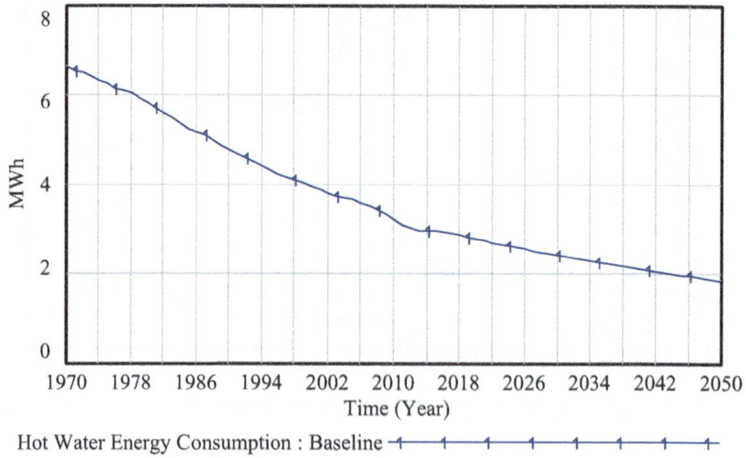

Fig. 6.16 Average hot water energy consumption per household

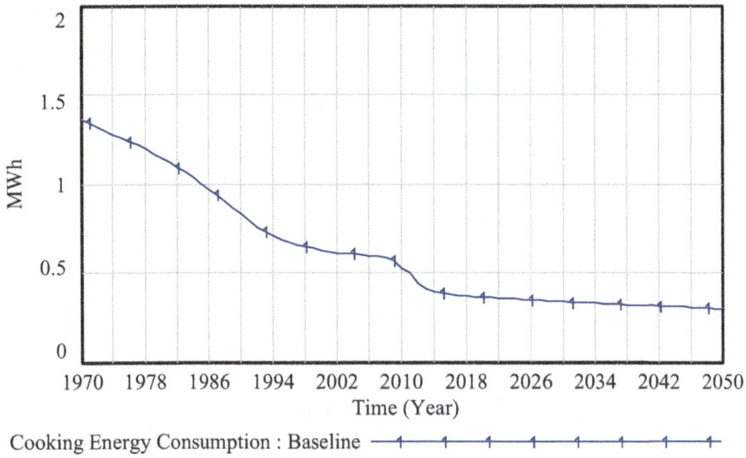

Fig. 6.17 Average cooking energy consumption per household

exhibited by the result shows a downward trend since 1970 until 2050. The behaviour as shown in Fig. 6.17 indicates that there is a steep slope from 1970 down to the 1990s. Afterwards, the downward trend seems to have levelled out for a short period since year 2000 except for a period between 2008 and 2016. The general downward pattern being witnessed could be due to a change in lifestyle where most families cook less or no longer cook at home and as a result eat at eateries, which consequently reduces the cooking energy consumption in homes. However, the trend levelling up is more pronounced around the year 2016 until 2050. This saw the slope of the trend of average household cooking energy to be gentler compared to the preceding years. The explanation that could be given to this trend may be the result of the size of households as shown in Sect. 6.3.1, which has been on a downward trend. As such, the Energy Saving Trust (EST), the DECC, and the Department of Environment, Food and Rural Affairs (DEFRA) [8] claimed that the cooking energy per head is higher in single-person households.

6.4.4 Behaviour Analysis of Lighting Energy Consumption

The simulation result for household lighting energy suggests that the energy consumption in this end-use has continued to be a small part of total household energy in general. The behaviour exhibited by the output of the simulation is shown in Fig. 6.18. The graph shows that the average lighting energy consumption per household followed an upward trajectory since 1970 until 2004 when it begins to decline gradually. This decrease could be due to government's policy of the Carbon Emissions Reduction Target (CERT) which ensures that energy-consuming incandescent bulbs are replaced in homes with energy-efficient ones [13].

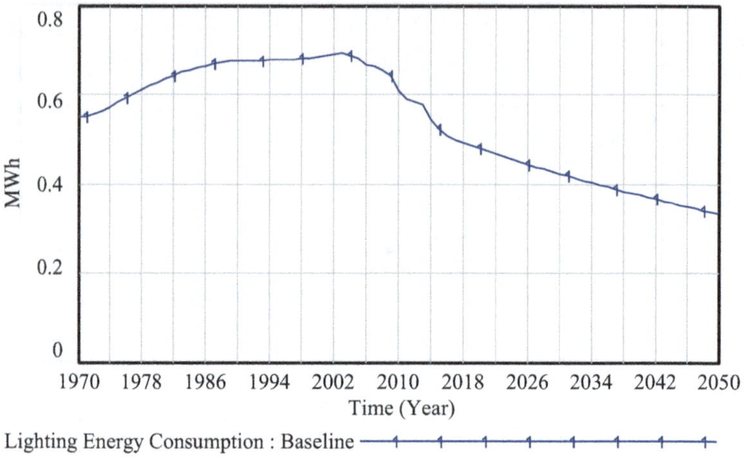

Fig. 6.18 Average lighting energy consumption per household

However, the simulation result suggests that the rate of decline of household lighting energy consumption would decrease as from 2016 as against the trend witnessed between 2004 and 2016. The reason for this, according to Oladokun and Odesola [13], could be attributed to an increase in the lighting points in homes, especially in the kitchens and bathrooms, which are even most times of higher specifications. Therefore, this is likely to reduce the rate of decline by offsetting the savings that would have been recorded should the trend of decline between 2004 and 2016 have been maintained.

6.4.5 Behaviour Analysis of Household Appliances' Energy Consumption

The result of the simulation run for household appliances' energy consumption is shown in Fig. 6.19. The insight from the result suggests that the household appliances energy use has been following an upward trajectory for the time horizon of the simulation. The pattern of behaviour, as shown in the result of Fig. 6.19, is seen to be consistent with the historical data provided by Palmer and Cooper [14].

There are some reasons that explain this pattern of behaviour according to the authors' conjecture. The first of these reasons could be that there is a surge in home/domestic electric gadgets which could be due to the fact that a number of homes have access to more disposable income. Acquisition of the gadgets may be due to change in the lifestyle of most occupants towards acquiring these latest electronic appliances. Furthermore, Oladokun and Odesola [13] argued that the owning of those gadgets alone may not result in an increase in the household appliances energy if they are not being used. However, the rate of using these gadgets has been

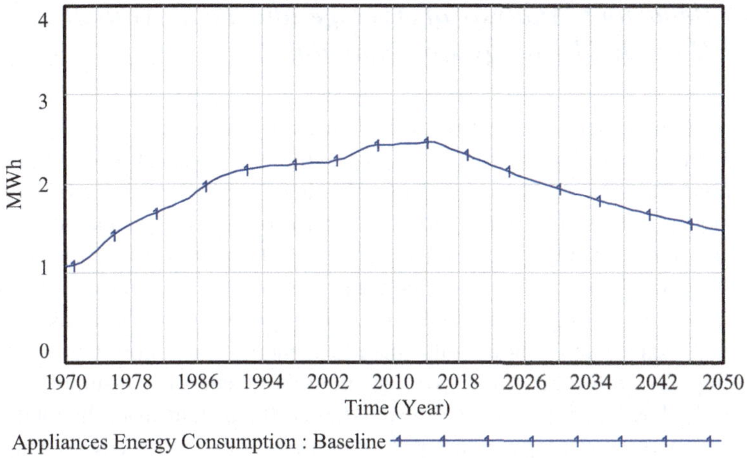

Fig. 6.19 Average appliances' energy consumption per household

on the upward trend, which invariably increases the amount of household appliances' energy. As explained previously, this could be attributed to the change in lifestyle of most occupants of dwellings. Additionally, using a room or rooms at home as offices as well as occupants changing to the use of energy-consuming appliances for some tasks or games that were traditionally completed manually could be responsible for this increase.

In addition to above, the insight from the simulation run as shown in Fig. 6.19 could be linked to the outcome of the study of EST et al. [8], which shows a surge in the use of some cold appliances such as freezers and large fridges. The study found that about one half of the total household appliances' energy consumption is attributed to cold appliances. Furthermore, the use of microwaves to defrost frozen food has also been reported to be on the increase. Combining these reasons suggests that the household appliances' energy has been on the increase. On the other hand, the simulation result as shown in Fig. 6.19 depicts that the household appliances' energy began to decline from 2016 following a gentle slope until 2050. This insight may be attributed to some positive breakthroughs in the research agendas targeting the energy efficiency of cold appliances. These research efforts would hopefully see the use of many more energy-efficient cold appliances in homes in the years to come, considering the fact that the major share of the household appliances' energy consumption comes from these appliances.

6.4.6 Behaviour Analysis of Average and Total Annual Household Energy Consumption

Considering the insights generated for the household energy consumption based on different end-uses as discussed in Sects. 6.4.1 to 6.4.5, Fig. 6.20 gives the average annual household energy consumption for the time horizon under consideration for the UK housing stock. The figure shows the profile of the average annual household energy consumption. One profound insight from the pattern of behaviour of the average annual household energy consumption is that it follows the pattern exhibited by the average household space heating energy consumption. This by implication means that the household space heat energy consumption is actually the moderating factor for the entire average household energy consumption in the whole of the UK housing stock. Correspondingly, the insight from the total annual household energy consumption that was computed for the entire UK housing stock as shown in Fig. 6.21 suggests a similar pattern of behaviour as the average household energy consumption. The computation of the total household energy consumption in the UK housing stock was done by multiplying the average annual household energy consumption by the number of households in the UK. The effect of the number of households in the model is noteworthy considering that the number of households in the UK has been on an upward trend over the years, most likely due to the conversion of some office buildings to homes. However, this may have overblown the total annual household energy consumption for the UK housing stock to some extent.

A further analysis regarding the household energy consumption based on end-uses for the entire UK housing stock is carried out for the simulation run. This

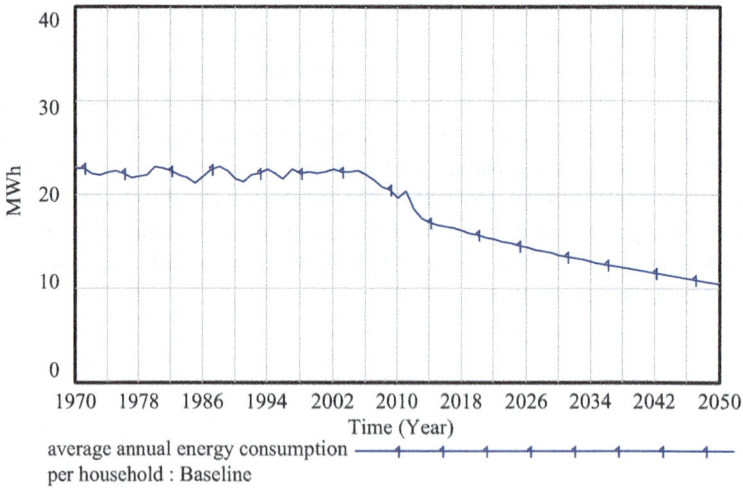

Fig. 6.20 Average annual energy consumption per household

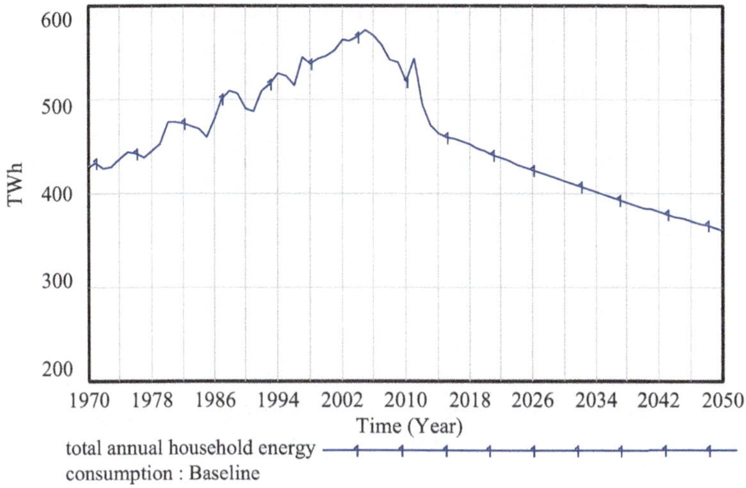

Fig. 6.21 Total annual energy consumption for the UK housing stock

analysis is conducted for the years 2020 and 2050 relative to the year 1990. This kind of analysis is mainly necessary to determine the extent to which the household energy consumption has been reduced or otherwise under the 'baseline' scenario assumptions for the model. Tables 6.2 and 6.3 illustrate the changes in household energy for the years 2020 and 2050 relative to the year 1990, respectively.

The results of this analysis for the year 2020 based on the 'baseline' scenario as shown in Table 6.2 suggest that the total household energy consumption for the entire UK housing stock is expected to witness a reduction of about 45.92 TWh of energy. This amount translates to about 9% reduction by 2020. Furthermore, the analysis within the period suggests that the space heating energy is expected to

Table 6.2 Change in household energy consumption by end-use based on 'baseline' scenario for the year 2020 relative to 1990

	Household energy consumption (1990) (TWh)	Household energy consumption (2020) (TWh)	Change in household energy consumption (TWh)[a]	Percentage change in household energy consumption (%)[a]
Space heating	300.92	276.88	−24.04	−7.99
Hot water	108.20	79.41	−28.79	−26.61
Cooking	18.88	10.52	−8.36	−44.28
Lighting	15.29	13.62	−1.67	−10.92
Appliances	47.93	64.87	+16.94	+35.34
Total	491.22	445.30	−45.92	−9.35

[a]Relative to 1990 base as enshrined in Climate Change Act of 2008

Table 6.3 Change in household energy consumption by end-use based on 'baseline' scenario for the year 2050 relative to 1990

	Household energy consumption (1990) (TWh)	Household energy consumption (2050) (TWh)	Change in household energy consumption (TWh)[a]	Percentage change in household energy consumption (%)[a]
Space heating	300.92	224.55	−76.37	−25.38
Hot water	108.20	63.73	−44.47	−41.10
Cooking	18.88	10.40	−8.48	−44.92
Lighting	15.29	11.44	−3.85	−25.18
Appliances	47.93	50.43	+2.5	+5.21
Total	491.22	360.55	−130.67	−26.60

[a]Relative to 1990 base as enshrined in Climate Change Act of 2008

witness a reduction of about 8%, hot water about 26%, cooking about 44%, and lighting about 11%. However, appliances' energy for the same period is expected to increase by approximately 35%.

For the year 2050, however, the analysis of results as shown in Table 6.3 indicates that the total household energy for the UK housing stock is expected to be reduced by approximately 27% relative to 1990 under the 'baseline' scenario assumptions. This percentage sees the total household energy less by 130.67 TWh when compared to 491.22 TWh in the base year 1990. Additionally, the results of the model suggest that the energy consumption due to space heating is expected to be reduced by 76.37 TWh, which translates to a reduction in energy of approximately 25% by 2050 relative to 1990 as base case. Also, the energy consumption attributable to hot water is anticipated to be reduced as well by 44.47 TWh, which amounts to a reduction of approximately 41%, again relative to 1990 base case. Correspondingly, the results of the model suggest that the energy consumption for cooking is expected to be marginally reduced by about 45% as against the 44% reduction envisaged for the year 2020. As expected, the model results indicate that the energy consumption for lighting is expected to be reduced as well by approximately 25%. Regarding the appliances' energy consumption, the model results suggest that this is anticipated to increase by approximately 5% relative to the base year 1990. However, this witnesses/indicates a reduction in consumption when compared to the results of the year 2020.

The changes expected to occur in energy consumption based on end-uses are based on the projection of the continuation of current trends in fabric insulation, energy efficiency, energy prices, and consumption behaviour. As fabric insulation ('insulation factor') and energy efficiency improve ('effect of energy efficiency standard improvement on dwelling energy efficiency'), a reduction in household energy consumption is anticipated. Also, standard consumption behaviour with a moderate rise in energy prices ('% increment on energy bills') is expected to lead to a reduction in household energy consumption by the year 2050 as the model output suggests. However, it must be emphasised that the results of the simulation run for the 'baseline' scenario indicate that the total number of UK households as well as the average internal temperature increases within this period. They now tend to increase the total household energy consumption. Within this period the occupants' thermal comfort also increases, as can be seen in Fig. 6.12 of Sect. 6.3.4 of this chapter. The implication of these would result in rebound effects as the majority of the accrued savings would have been expended on obtaining improved comfort.

6.5 Behaviour Analysis of Household Carbon Emissions Module

This section of the book discusses the main results of the simulation run under the household carbon emissions module. The results are discussed for carbon emissions based on end-uses of energy. That is, the results are presented based on carbon emissions due to space heating, water heating, cooking, lighting, and appliances usage. The section also discusses the results of the average and total annual household carbon emissions of the entire UK housing stock.

6.5.1 Behaviour Analysis of Carbon Emissions Due to Space Heating Energy

Figure 6.22 shows the behaviour over time of the average space heating carbon emissions per household in the UK. The result reveals that carbon emissions due to space heating have been on a downward trend since 1970. The insight from the simulation indicates that this downward trend will continue until the year 2050. The model suggests that the average carbon emissions due to space heating per household are likely to decrease from about six tonnes in 1970 to about one tonne by the year 2050. The decrease in the emissions' profile over the years may be due to the fact that there are improvements in clean energy technology.

6.5.2 Behaviour Analysis of Carbon Emissions Due to Hot Water Energy Usage

The output of the simulation run for carbon emissions due to hot water usage is shown in Fig. 6.23. Similar to carbon emissions due to space heating, the result is shown for the average emissions per household. The simulation result suggests almost the same pattern of behaviour as depicted for the hot water energy use in

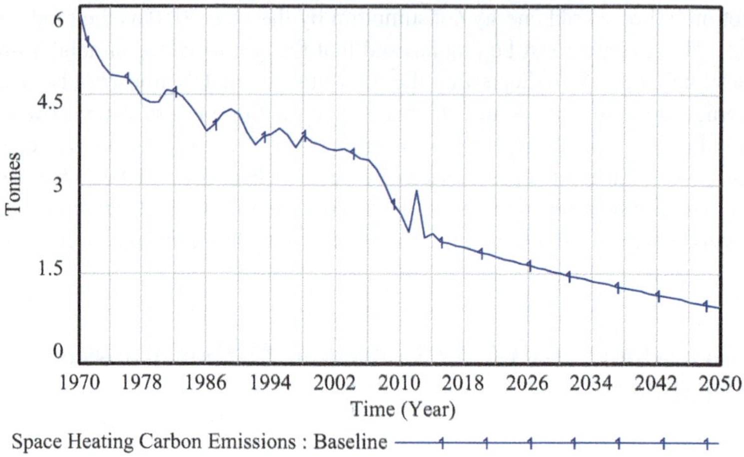

Fig. 6.22 Average space heating carbon emissions per household

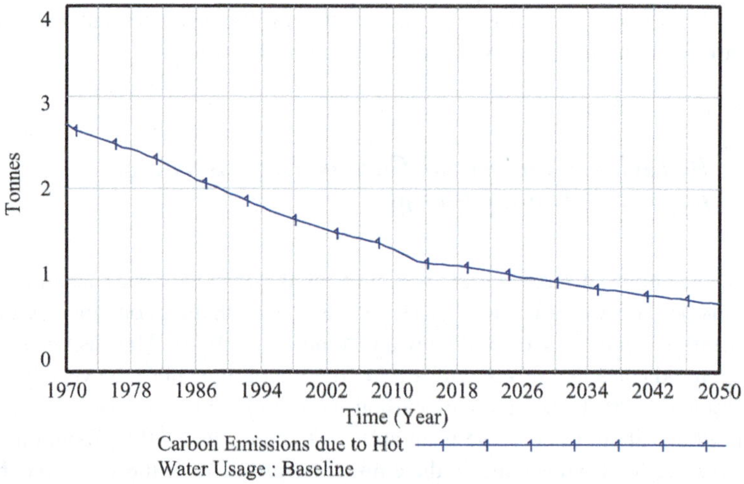

Fig. 6.23 Average carbon emissions due to hot water usage per household

Sect. 6.4.2. The insight from the output indicates that the average carbon footprints of each household are on a downward trajectory. This by implication means that carbon emissions due to hot water usage in UK dwellings are about one half of the average emissions attributable to space heating. As such, the amount of carbon emissions due to water heating is still high, though remarkable reductions have been noticed over the years. The reductions may be attributed to the fact that hot water energy usage has also shown on a downward trend over the same time horizon owing to the reasons advanced in Sect. 6.4.2 regarding reductions in hot water energy usage. Other reasons could be either that the improvements in the clean energy technology are sustained or more improvements are made.

6.5.3 Behaviour Analysis of Carbon Emissions Due to Cooking Energy

The behaviour analysis for the simulation output as generated for the average carbon emissions due to cooking energy per household is shown in Fig. 6.24. The result indicates the pattern of behaviour expected from the carbon emissions due to cooking energy in dwellings. Generally, the trend shown is in no way different from the pattern exhibited by the result for the cooking energy consumption shown previously in Fig. 6.17. The profile of average carbon emissions due to cooking energy suggests a downward trend from the year 1970 to 2050 apart from the fact that there is a steep slope from 1970 down to around 1994. This then follows a kind of 's' curve profile before levelling out towards the year 2050. The main reason for this behaviour may be the influence of the energy consumption pattern for the cooking end-use of energy. This may be based on the reduction in the household size as well as eating patterns as advanced by the EST, the DECC, and the DEFRA [8].

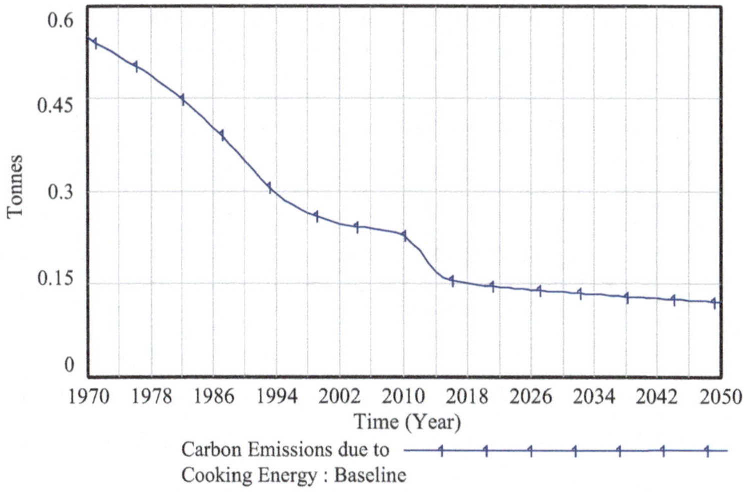

Fig. 6.24 Average carbon emissions due to cooking energy per household

6.5.4 Behaviour Analysis of Carbon Emissions Due to Lighting Energy

For the behaviour of the simulation output for the carbon emissions due to lighting energy, the trend is shown in Fig. 6.25. Generally, the carbon emissions due to lighting energy are a small fraction of the total carbon emissions in homes. The pattern of behaviour as depicted in Fig. 6.25 replicates almost the same trajectory witnessed under the household energy consumption for lighting. That is, the graph reveals a growth in carbon emissions due to lighting energy from 1970 up to around 2014 before declining gradually until 2050. The reason may be connected to the government's policy on energy efficiency targets on lighting bulbs as reported by Oladokun and Odesola [13].

6.5.5 Behaviour Analysis of Carbon Emissions Due to Household Appliances' Energy

Figure 6.26 shows the profile of household carbon emissions due to appliances' energy consumption based on the simulation run. The insight from the result suggests nearly the same pattern of output as witnessed under the household appliances' energy use. The pattern of behaviour shows that the average household carbon emissions due to appliances' energy have been on an upward trajectory since 1970 and up to around 2016 before an event overturn. Around 2016, the average household carbon emissions due to appliance energy started to decline and

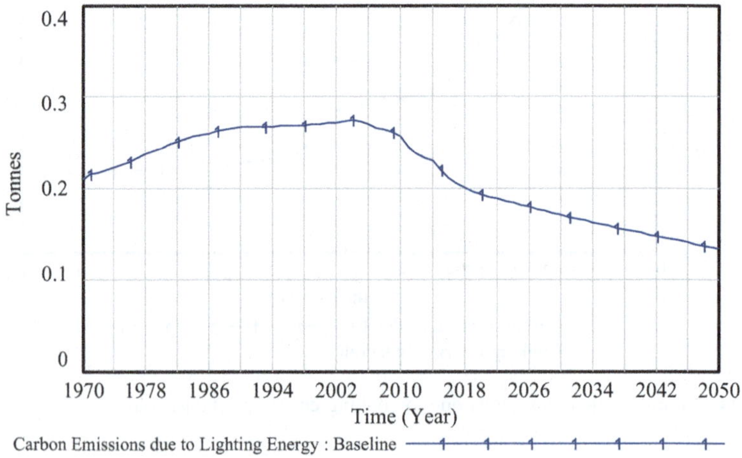

Fig. 6.25 Average carbon emissions due to lighting energy per household

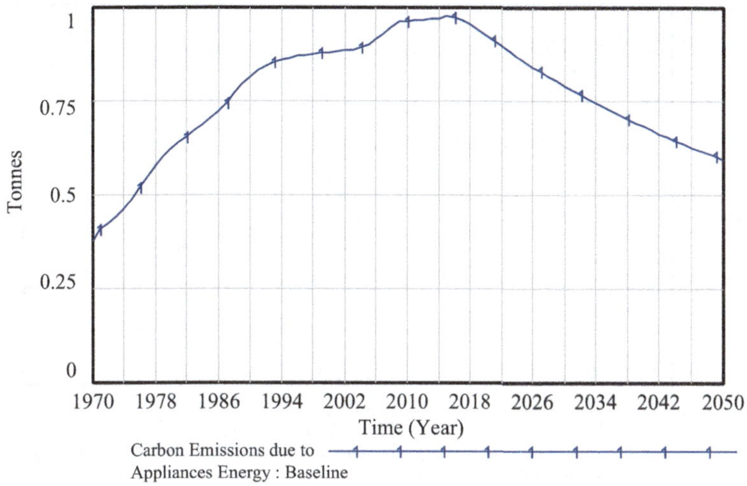

Fig. 6.26 Average carbon emissions due to appliances energy per household

the output of the simulation run dictates that it will continue to do so until 2050. The reasons that can be advanced for this pattern of behaviour are not far-fetched as mentioned previously in Sect. 6.4.5.

6.5.6 Behaviour Analysis of Average and Total Annual Household Carbon Emissions

The preceding Sects. 6.5.1–6.5.5 described the simulation results of the average household carbon emissions by end-use. This subsection of the book explains the pattern of behaviour expected for the household carbon emissions in terms of the average annual and total household carbon emissions. The results of the simulation for the average annual and total household carbon emissions are shown in Figs. 6.27 and 6.28, respectively. These results are profound based on the pattern of behaviours they exhibit. The trends shown in the two figures may be due to the fact that carbon emissions are a result of energy consumption. However, the dominant type of energy consumed by householders would go a long way in moderating household carbon emissions. When assessing the average annual carbon emissions per household (Fig. 6.27) and total annual household carbon emissions (Fig. 6.28), it can be noted that carbon emissions have been on a downward trajectory since 1970. That is, the average annual carbon emissions per household have fallen significantly since 1970 and the model projects that the trend will be sustained till 2050 based on the carbon reductions' agenda of the government. The output is similar to the trend witnessed in historical data [14] as the trend (Figs. 6.27 and

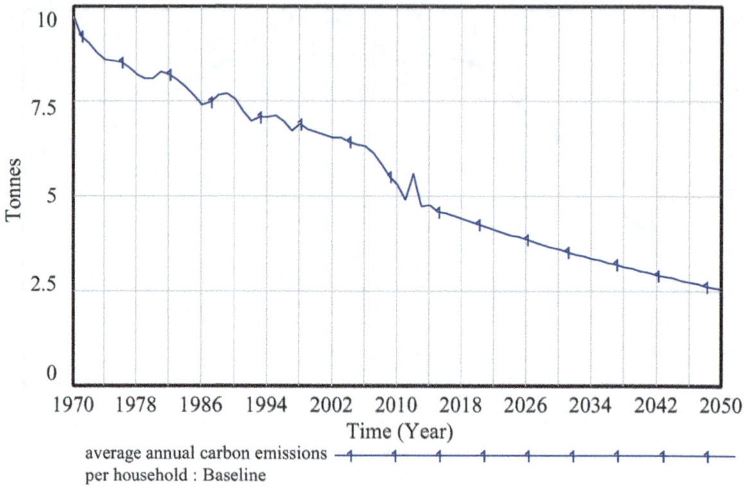

Fig. 6.27 Graph of average annual household carbon emissions

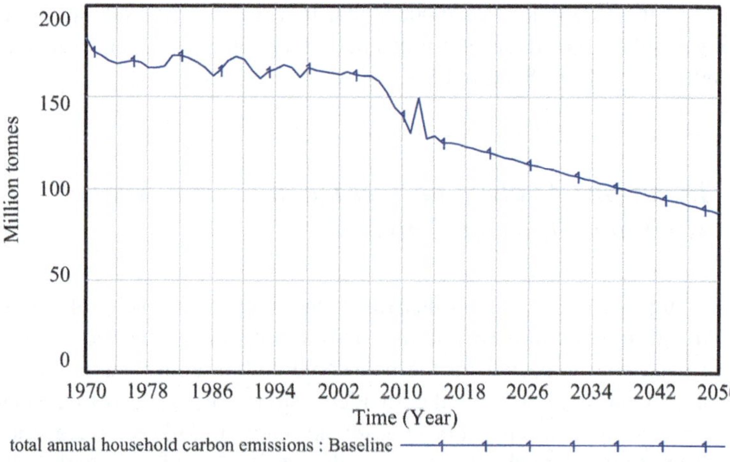

Fig. 6.28 Graph of total household carbon emissions

6.28) tends to be 'lumpy' with troughs and peaks that correspond to mild and severe weather conditions.

It is equally important to conduct a further analysis of results in order to reveal additional insights as well as to determine the extent to which the carbon emissions reductions' target is achieved for the baseline scenario. Tables 6.4 and 6.5 illustrate the changes or reductions expected in household carbon emissions for the years 2020 and 2050 relative to the year 1990. The Climate Change Act of 2008 in the

Table 6.4 Change in household carbon emissions by end-use based on 'baseline' scenario for the year 2020 relative to 1990

	Household carbon emissions (1990) (million tonnes of CO_2)	Household carbon emissions (2020) (million tonnes of CO_2)	Change in household carbon emissions (million tonnes of CO_2)[a]	Percentage change in carbon emissions (%)[a]
Space heating	94.47	53.19	−41.28	−43.70
Hot water	44.15	32.09	−12.06	−27.32
Cooking	7.93	4.21	−3.72	−46.91
Lighting	6.04	5.50	−0.54	−8.94
Appliances	18.43	26.29	+7.86	+42.65
Total	171.01	121.28	−49.73	−29.08

[a]Relative to 1990 base as enshrined in Climate Change Act of 2008

Table 6.5 Change in household carbon emissions by end-use based on 'baseline' scenario for the year 2050 relative to 1990

	Household carbon emissions (1990) (million tonnes of CO_2)	Household carbon emissions (2050) (million tonnes of CO_2)	Change in household carbon emissions (million tonnes of CO_2)[a]	Percentage change in carbon emissions (%)[a]
Space heating	94.47	32.46	−62.01	−65.64
Hot water	44.15	25.71	−18.44	−41.77
Cooking	7.93	4.16	−3.77	−47.54
Lighting	6.04	4.61	−1.43	−23.68
Appliances	18.43	20.35	+1.92	+10.42
Total	171.01	87.28	−83.73	−48.96

[a]Relative to 1990 base as enshrined in Climate Change Act of 2008

UK stipulates a carbon emissions reductions' target of 34 and 80% relative to 1990 level by the years 2020 and 2050, respectively.

To this end, the simulation results in Table 6.4 suggest that the carbon emissions ascribable to the UK households are expected to decline by 49.73 million tonnes of carbon emissions by the year 2020. This amount represents a reduction in carbon emissions of approximately 29%. The implication of this result is that under the 'baseline' scenario it unlikely to meet the target reduction of 34% as enshrined in the Climate Change Act of 2008. A further analysis based on end-uses reveals that the greatest reductions are expected to happen in space heating, which is anticipated to witness a 41.28 million tonnes of carbon emissions' reduction by the year 2020.

The analysis conducted for the year 2050 reveals that the carbon emissions attributable to the UK housing stock are expected to decline by up to about 83.73

million tonnes of carbon emissions. This amount represents a reduction in carbon emissions of approximately 49% by the middle of this century. Similarly, the implication of this result under the 'baseline' scenario suggests that it unlikely to meet the target reduction of 80% as enshrined in the Climate Change Act of 2008. An additional analysis based on end-uses shows that the majority of the reductions expected in carbon emissions are to occur in space heating. A reduction of approximately 62 million tonnes of carbon emissions is expected by the year 2050.

6.6 Model Testing and Validation

In the system dynamics approach, the model testing and validation is regarded as important stage. This section of the book reports on the model testing and validation process completed for the simulation run. The section first discusses the SD validation tests that can be performed. Subsequently, the section discusses the results of the validation tests performed in terms of structure-oriented and behaviour pattern tests conducted.

6.6.1 Model Validation Tests

The simulation run for this research book was implemented on Vensim DSS for Windows software. Vensim is one of the SD modelling tools that are commonly used to build, simulate, and analyse SD models. Researchers acknowledge model testing and validation (MTaV) as an important aspect of any model-based methodology such as SD [3, 15]. As such, MTaV is a crucial step that is not to be disregarded. It is significant in the sense that the validity of results emanating from the model depends heavily on the validity of the model itself. MTaV is the process of testing the soundness and correctness of the construction of the model while establishing confidence in the usefulness of the model [6, 7]. Hence, this MTaV exercise proves the credibility of the outputs from the model and ascertains that the results accurately represent reality. Testing the model actually means validating it. However, some researchers argue that MTaV is a controversial issue [3] because there is no single approach that would allow the modellers to ascertain that their models have been validated. Further to this controversy, Sterman [18] contends that complete model validation is practically impossible and as such, more emphasis needs to be placed on model testing in order to build confidence that the model is adequate for the intended purpose.

To this end, there are a number of tests to assess the validity of SD models. These are generally divided into structure-oriented and behaviour pattern tests [2, 3, 9, 10, 16, 18]. The tests include and are not limited to (1) structure-oriented tests—boundary adequacy, structure assessment, dimensional consistency, parameter assessment, extreme conditions, and integration error, and (2) behaviour pattern tests—behaviour reproduction, behaviour anomaly, family member, surprise behaviour, sensitivity analysis, and system improvement. The purpose of each of these tests and tools or procedures is illustrated in Table 6.6, as adapted from the work of Sterman [18].

6.6.2 Structure-Oriented Tests

The purpose of structure-oriented tests is to ensure that the simulation results are able to capture the real system being replicated. According to Ranganath and Rodrigue [15], the tests are to build confidence and ensure that the simulation run is appropriate for the target audience. Additionally, the tests focus on the suitability of the level of aggregation and determine whether the basic physical laws are strictly adhered to regarding the parameters utilised in the model. In this research book, the developed model is subjected to a number of structure-oriented tests for the purpose of validation. The following are the tests performed for the structure-oriented tests:

Boundary adequacy test

As shown in Table 6.6, the boundary adequacy tests assess the appropriateness of the model boundary to capture the problem under investigation. The model boundary charts for this model are shown in Table 5.2 in Chap. 5 of this book. The model boundary chart is one of the useful tools that can be used to conduct boundary adequacy tests. Further to the model boundary chart, the model CLDs of different modules went through different series of validation exercises, which include qualitative validation through a series of interviews held with experts and practitioners on issues relating to household energy. The feedback from these interviews indicates that the study captures important variables relating to HECE. However, some of the interviewees are of the opinion that there is more room for improvement through making some of the exogenous variables such as occupants' behaviour in the model endogenous as well as expanding the boundary of the model by including some of the excluded variables in the upgraded version of the model.

Structure assessment test

The structure assessment tests whether the model structure is consistent with the relevant descriptive knowledge of the system and conforms to basic physical laws [18]. It also tests whether the level of aggregation of the model is appropriate as tested under the model boundary adequacy tests. In this research book, the authors ensure that the structure of the developed model considers all the real-life issues that are consistent with the relevant descriptive knowledge on the subject. This is done at both the model conceptualisation and complete model validation stages.

Table 6.6 SD validation tests

Test	Purpose of test	Recommended tools and procedures
A. Structure validity		
1. Boundary adequacy	Are the important concepts for addressing the problem endogenous to the model? Does the behaviour of the model change significantly when boundary assumptions are relaxed? Do policy recommendations change when the boundary is extended?	Use model boundary charts, subsystem diagrams, causal diagrams, stock and flow maps, and direct inspection of model equations Use interviews, workshops to solicit expert opinion, archival materials, review of literature, direct inspection/participation in system processes, and the like Modify model to include plausible additional structure; make constants and exogenous variables endogenous, then repeat sensitivity and policy analysis
2. Structure assessment	Is the model structure consistent with relevant descriptive knowledge of the system? Is the level of aggregation appropriate? Does the model conform to basic physical laws such as conservation laws? Do the decision rules capture the behaviour of the actors in the system?	Use policy structure diagrams, causal diagrams, stock and flow maps, and direct inspection of model equations Use interviews, workshops to solicit expert opinion, archival materials, direct inspection or participation in system processes, as in (1) above Conduct partial model tests of the intended rationality of decision rules Conduct laboratory experiments to elicit mental models and decision rules of system participants Develop disaggregate submodels and compare behaviour to aggregate formulations Disaggregate suspect structures, then repeat sensitivity and policy analysis
3. Dimensional consistency	Is each equation dimensionally consistent without the use of parameters having no real world meaning?	Use dimensional analysis software Inspect model equations for suspect parameters

(continued)

Table 6.6 (continued)

Test	Purpose of test	Recommended tools and procedures
4. Parameter assessment	Are the parameter values consistent with relevant descriptive and numerical knowledge of the system? Do all parameters have real world counterparts?	Use statistical methods to estimate parameters (wide range of methods available) Use partial model tests to calibrate subsystems Use judgemental methods based on interviews, expert opinion, focus groups, archival materials, direct experience, etc Develop disaggregate submodels to estimate relationships for use in more aggregate models
5. Extreme conditions	Does each equation make sense even when its inputs take on extreme values? Does the model respond plausibly when subjected to extreme policies, shocks, and parameters?	Inspect each equation Test response to extreme values of each input, alone and in combination Subject model to large shocks and extreme conditions. Implement tests that examine conformance to basic physical laws
6. Integration error	Are the results sensitive to the choice of time step or numerical integration method?	Cut the time step in half and test for changes in behaviour Use different integration methods and test for changes in behaviour
B. Behaviour validity		
7. Behaviour reproduction	Does the model reproduce the behaviour of interest in the system (qualitatively and quantitatively)? Does it endogenously generate the symptoms of difficulty motivating the study? Does the model generate the various modes of behaviour observed in the real system? Do the frequencies and phase relationships among the variables match the data?	Compute statistical measures of correspondence between model and data: descriptive statistics (e.g. R^2); time domain methods (e.g. autocorrelation functions); frequency domain methods (e.g. spectral analysis); many others Compare model output and data qualitatively, including modes of behaviour, shape of variables, asymmetries, relative amplitudes and phasing, unusual events Examine response of model to test inputs, shocks, and noise
8. Behaviour anomaly	Do anomalous behaviours result when assumptions of the model are changed or deleted?	Zero out key effects (loop knockout analysis) Replace equilibrium assumptions with disequilibrium structures

(continued)

Table 6.6 (continued)

Test	Purpose of test	Recommended tools and procedures
9. Family member	Can the model generate the behaviour observed in other instances of the same system?	Calibrate the model to the widest possible range of related systems
10. Surprise behaviour	Does the model generate previously unobserved or unrecognised behaviour? Does the model successfully anticipate the response of the system to novel conditions?	Keep accurate, complete, and dated records of model simulations. Use model to simulate likely future behaviour of system Resolve all discrepancies between model behaviour and your understanding of the real system Document participant and client mental models prior to the start of the modelling effort
11. Sensitivity analysis	*Numerical sensitivity:* Do the numerical values change significantly…. *Behavioural sensitivity:* Do the modes of behaviour generated by the model change significantly…. *Policy sensitivity:* Do the policy implications change significantly…. ….when assumptions about parameters, boundary, and aggregation are varied over the plausible range of uncertainty?	Perform univariate and multivariate sensitivity analysis Use analytic methods (linearisation, local and global stability analysis, and the like) Conduct model boundary and aggregation tests listed in (1) and (2) above Use optimisation methods to find the best parameters and policies Use optimisation method to find parameter combinations that generate implausible results or reverse policy outcomes
12. System improvement	Did the modelling process help change the system for the better?	Design instruments in advance to assess the impact of the modelling process on mental models, behaviour, and outcomes Design controlled experiments with treatment and control groups, random assignment, pre-intervention and post-intervention assessment, and the like

Adapted from Sterman [18]

At the model validation stage, the results of the interviews indicate that both the energy and SD experts interviewed are satisfied with the structure of the model in terms of its description of the relevant knowledge in the subject. One of the interviewees reports that '*...the structure of the model makes it easy to follow and is simple to understand even if you are not familiar with energy issues or system dynamics*'. Another interviewee expresses the view that '*...the structure of the model demonstrates how a large number of variables are interrelated in a logical* *manner*'. When asked to comment generally on the structure of the model, another interviewee notes that '*...this is a very well developed model which appears to represent a very impressive body of work*'.

Additionally, both the authors and the system dynamicists qualitatively interviewed inspected some of the model equations in order to assess whether they are conformable to and consistent with the basic physical laws. The outcome of the validation indicates that all the equations inspected are conformable to the basic physical laws and do make sense. Also, the reports of the interviews conducted with the energy experts suggest that the model includes all the significant variables and the level of aggregation is consistent with the target audience for the model.

Dimensional consistency test

The developed model was subjected to the dimensional consistency test as well. The dimensional consistency test assesses the model equations for dimensional consistency and checks whether all the units and values attributed to the model parameters are consistent with relevant understanding of the system under investigation [18]. In order to ensure that the model is validated accordingly, the dimensional analysis tool within the Vensim software was used to conduct this test. In Vensim, the software automatically checks the dimensions of all the variables and the equations in the model in order to ascertain whether they are consequently balanced. For the model in this book, the dimensional analysis tool was invoked and all the units of the model variables and equations were verified and balanced accordingly.

Parameter assessment test

According to Sterman [18], the parameter assessment test evaluates the model parameters and checks whether all their values are consistent with relevant descriptive and numerical noesis of the system. This is consistent with the argument of Ranganath and Rodrigues [15] that the numerical values of parameters should have real system equivalents. In order to ensure that the parameter assessment of the model variables is adequately evaluated, Sterman [18] suggests the tools and procedures for achieving this as shown in Table 6.6. Among the ways suggested are the statistical and judgemental methods. The values of the majority of parameters in the model are taken from relevant published data sources. For example, the parameters (*solar flux, solar transmittance factor for glazing, frame factor, and the like.*) under the dwelling internal heat module are extracted from the SAP data [4].

Extreme conditions test

The extreme conditions test evaluates how sound and robust the model is based on its response to the variables subjected to extreme values. This test then assesses the model equations and checks whether they still make any sense when subjected to extreme conditions. In order to test and validate the model against this structure-oriented test, the model was subjugated to extreme values of some parameters. For example, the model was submitted to extreme values of 'insulation factor' and '% increment of energy bills', which are varied for 0 and 100%. The model results indicate that the behaviour of the output still makes sense without any plausible or irrational response in terms of the values of outputs as shown in Figs. 6.29, 6.30, 6.31, and 6.32.

Integration error test

It is necessary to test the robustness of the model using the integration error tests. Sterman [18] argues that the integration error tests the sensitivity of the model results to the choice of *Time Step* and/or numerical integration methods employed in the simulation. He recommends cutting the *Time Step* used in the simulation is half in order to check whether there are changes in the behaviour of the model outputs. Similarly, Sterman [18] submits that changes in model behaviour need to be checked and tested against different integration methods.

In this research book, both the changes attributed to the choice of *Time Step* and integration method tests were conducted. The integration error test was conducted by first splitting the *Time Step* of one (*Time Step* = *1*) used for the simulation into two (i.e. *Time Step* = *0.5*) and then running the simulation again. The final outputs (in terms of household energy consumption and carbon emissions) of the models were examined to check whether there are changes in their behaviour. In order to

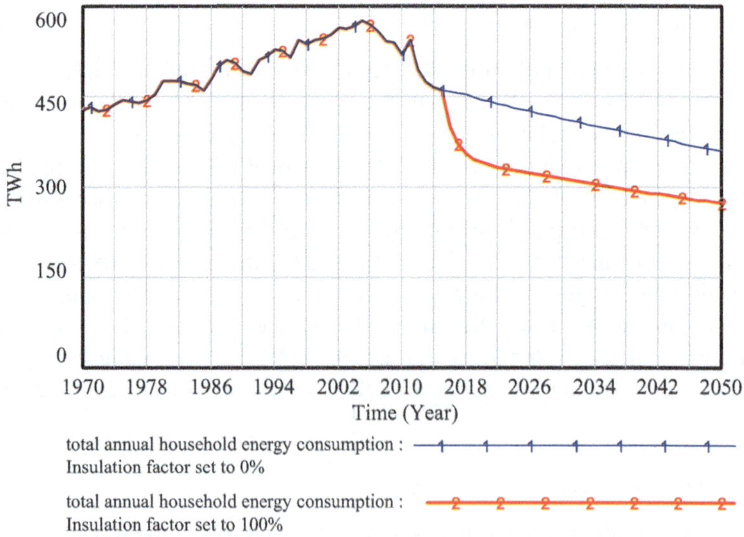

Fig. 6.29 Total annual household energy consumption under 'insulation factor' set to 0 and 100%

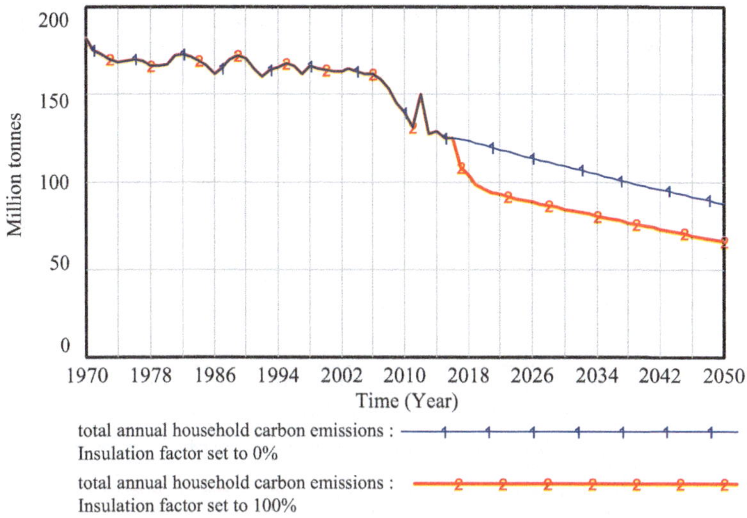

Fig. 6.30 Total annual household carbon emissions under 'insulation factor' set to 0 and 100%

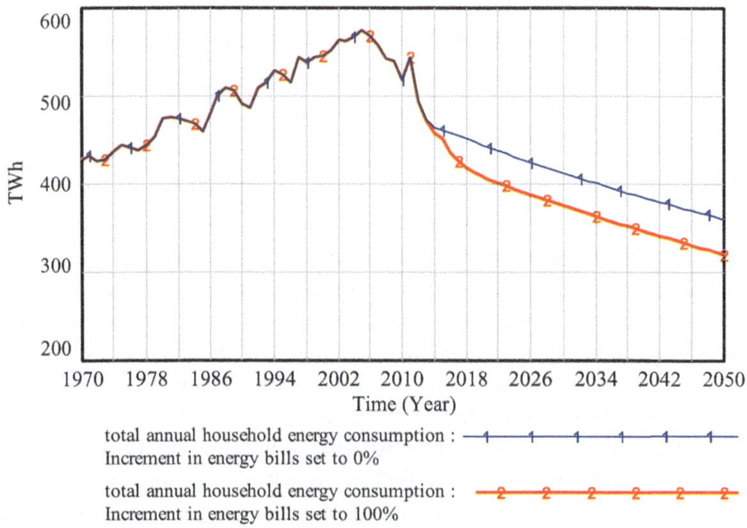

Fig. 6.31 Total annual household energy consumption under 'increment in energy bills' set to 0 and 100%

test whether there are changes, a hypothesis was set up. The null hypothesis (H_0) signifies that there is no statistically significant difference between the means of the model outputs for *Time Step = 1* and *Time Step = 0.5*, while the alternate hypothesis (H_1) signifies that there is a statistically significant difference between

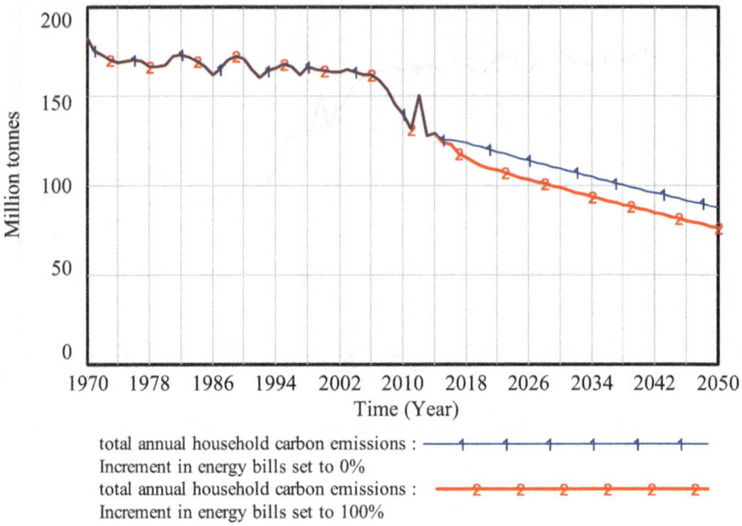

Fig. 6.32 Total annual household carbon emissions under 'increment in energy bills' set to 0 and 100%

the two means. Mathematically the hypothesis was set up as shown in the equations below:

$$H_0 : \mu_{i\,Time\,Step=1} - \mu_{i\,Time\,Step=0.5} = 0 \quad (6.1)$$

$$H_0 : \mu_{i\,Time\,Step=1} - \mu_{i\,Time\,Step=0.5} \neq 0 \quad (6.2)$$

where μ_i indicates the mean of variable of interest in the model.

A paired-sample t-test was conducted for ten different variables in the model, which include energy consumption and carbon emissions for space heating, hot water, cooking, lighting, and appliances for the UK housing stock as shown in Table 6.7. In order to take a decision regarding the hypothesis, the significance value (p value) is compared to the significance level ($\alpha = 0.05$), and based on these two values, the null hypothesis is either rejected or not rejected. If the p value is less than the significance level, the null hypothesis is rejected (i.e. p value $< \alpha$, reject the null); otherwise the null hypothesis is not rejected. In this case, the t statistics computed for the ten paired variables of interest as shown in Table 6.7 reveal that the p value for all the ten paired variables is greater than the value of significance level. This means that the p values of 0.527, 0.181, 0.251, 0.259, 0.779, 0.495, 0.418, 0.567, and 0.320 are greater than the α-value of 0.05. Therefore, the null hypothesis is accepted and the alternative hypothesis rejected. This by implication means that there is no statistically significant difference in the means of output of variables of interest shown in Table 6.7. That is, the results indicate that there are no changes in the behaviour of those selected outputs.

Table 6.7 Paired-sample *t*-test for *Time Step* changes

		Paired differences			*t*	df	Sig. (2-tailed)
		Mean	Std. deviation	Std. error mean			
Pair 1	SHEC_1–SHEC_0.5	0.00827	0.11729	0.01303	0.635	80	0.527
Pair 2	HWEC_1–HWEC_0.5	0.00111	0.00742	0.00082	1.348	80	0.181
Pair 3	CEC_1–CEC_0.5	−0.00049	0.00384	0.00043	−1.157	80	0.251
Pair 4	LEC_1–LEC_0.5	−0.00037	0.00293	0.00033	−1.136	80	0.259
Pair 5	AEC_1–AEC_0.5	0.00049	0.00384	0.00043	1.157	80	0.251
Pair 6	SHCE_1–SHCE_0.5	0.00222	0.07101	0.00789	0.282	80	0.779
Pair 7	HWCE_1–HWCE_0.5	0.00037	0.00486	0.00054	0.686	80	0.495
Pair 8	CCE_1–CCE_0.5	−0.00025	0.00273	0.00030	−0.815	80	0.418
Pair 9	LCE_1–LCE_0.5	0.00012	0.00193	0.00021	0.575	80	0.567
Pair 10	ACE_1–ACE_0.5	0.00037	0.00333	0.00037	1.000	80	0.320

SHEC space heating energy consumption, *HWEC* hot water energy consumption, *CEC* cooking energy consumption, *LEC* lighting energy consumption, *AEC* appliances energy consumption, *SHCE* carbon emissions due to space heating energy, *HWCE* carbon emissions due to hot water energy, *CCE* carbon emissions due to cooking energy, *LCE* carbon emissions due to lighting energy, *ACE* carbon emissions due to appliances energy, *Sig.* significance, *Std.* standard, *df* degree of freedom

In addition to cutting the *Time Step* in half with the use of the *Euler* numerical integration method, the model is simulated using four different integration methods. These integration methods are (1) fixed second-order Runge–Kutta (*fRK2*), (2) auto second-order Runge–Kutta (*aRK2*), (3) fixed fourth-order Runge–Kutta (*fRK4*), and (4) auto fourth-order Runge–Kutta (*aRK4*). As for the hypothesis testing above under the *Time Step* splitting, hypotheses were again set up as follows:

1. To test whether there is a statistically significant difference between the results of the simulation performed by *Euler* and *fRK2* numerical integration methods, the hypotheses in Eqs. 6.3 and 6.4 are formulated.

$$H_0 : \mu_{i\,Euler} - \mu_{i\,fRK2} = 0 \tag{6.3}$$

$$H_1 : \mu_{i\,Euler} - \mu_{i\,fRK2} \neq 0 \tag{6.4}$$

2. To test whether there is a statistically significant difference between the results of the simulation performed by *Euler* and *aRK2* numerical integration methods, the hypothesis in Eqs. 6.5 and 6.6 is formulated.

$$H_0 : \mu_{i\,Euler} - \mu_{i\,aRK2} = 0 \tag{6.5}$$

$$H_1 : \mu_{i\,Euler} - \mu_{i\,aRK2} \neq 0 \tag{6.6}$$

3. To test whether there is a statistically significant difference between the results of the simulation performed by *Euler* and *fRK4* numerical integration methods, the hypothesis in Eqs. 6.7 and 6.8 is formulated.

$$H_0 : \mu_{i\,Euler} - \mu_{i\,fRK4} = 0 \tag{6.7}$$

$$H_1 : \mu_{i\,Euler} - \mu_{i\,fRK4} \neq 0 \tag{6.8}$$

4. To test whether there is a statistically significant difference between the results of the simulation performed by *Euler* and *aRK4* numerical integration methods, the hypothesis in Eqs. 6.9 and 6.10 is formulated.

$$H_0 : \mu_{i\,Euler} - \mu_{i\,aRK4} = 0 \tag{6.9}$$

$$H_1 : \mu_{i\,Euler} - \mu_{i\,aRK4} \neq 0 \tag{6.10}$$

Again, a paired-sample t-test was carried out for energy consumption and carbon emissions of space heating, hot water, cooking, lighting, and appliances for the entire UK housing stock as shown in Table 6.8. The same decision rules as for the *Time Step* splitting were used to test whether there is any statistically significant difference between the behaviour of simulation outputs performed by *Euler* and four other integration methods. As before, the t statistics computed for the ten paired variables of interest as shown in Table 6.8 for each of *fRK2*, *aRK2*, *fRK4*, and *aRK4* show that the p value for all the ten paired variables is greater than the value of significance level for all the four integration methods (Table 6.8). Therefore, the null hypothesis is accepted and the alternative hypothesis rejected for all of them. This by implication means that it is safe to say that there is no statistical or significant difference in the means of output of variables of interest shown in Table 6.8. That is, there are no significant changes in the behaviour of these variables of interest.

6.6.3 Behaviour Pattern Tests

The main reason for conducting the behaviour pattern tests is to ensure that the simulation results are consistent with the behaviour patterns of the historical time series data of the variables in the real system under investigation [15]. A model is

Table 6.8 Paired-sample *t*-test for integration method changes

	Euler-fixed RK2			Euler-auto RK2			Euler-fixed RK4			Euler-auto RK4		
	t	df	Sig. (2-tailed)	*t*	df	Sig. (2-tailed)	*t*	df	Sig. (2-tailed)	*t*	df	Sig. (2-tailed)
Household energy consumption												
Space heating	−1.671	80	0.099	−1.556	80	0.124	−1.707	80	0.092	−1.523	80	0.132
Hot water	−1.937	80	0.056	−0.757	80	0.451	−1.472	80	0.145	−1.625	80	0.108
Cooking	−1.811	80	0.074	−1.114	80	0.269	−1.944	80	0.055	−1.114	80	0.269
Lighting	−1.401	80	0.165	−1.029	80	0.307	−1.865	80	0.066	−1.029	80	0.307
Appliances	1.044	80	0.300	1.857	80	0.067	1.668	80	0.099	1.811	80	0.074
Household carbon emissions												
Space heating	−1.839	80	0.070	−1.959	80	0.054	−1.039	80	0.302	−1.961	80	0.053
Hot water	−0.827	80	0.411	−1.862	80	0.066	−1.862	80	0.066	−1.862	80	0.066
Cooking	−1.089	80	0.279	−0.438	80	0.663	1.389	80	0.169	−0.438	80	0.663
Lighting	−0.630	80	0.530	−1.598	80	0.114	−0.630	80	0.530	−0.630	80	0.530
Appliances	0.883	80	0.380	0.961	80	0.339	0.847	80	0.399	0.961	80	0.339

therefore considered validated behaviourally if the results of the simulation performed give similar behavioural patterns when compared with the behaviour patterns observed in the time series data of the real system [18]. The historical time series of key variables has previously been shown in Chap. 5 (Sect. 5.4 of this book). As such, the behaviour pattern validation for the simulation run in this book is achieved by comparing the pattern of behaviour of the baseline simulation run with the historical time series data (reference modes) and some previously developed models using another approach [11]. Examples of the behaviour pattern tests conducted are as follows:

Behaviour anomaly test

According to Ranganath and Rodrigues [15], the behaviour anomaly test assesses the behaviour shown by the model and checks whether its output conflicts in any way with the real system behaviour. In addition, Sterman [18] submits that the behaviour anomaly test evaluates how implausible behaviour arises should the assumptions made in the model be altered. In order to conduct this test in this research, a loop knockout analysis was carried out on one of the loops in the occupants' thermal comfort module and its effect was assessed on the output of the model (Fig. 6.33). The results of the behaviour anomaly test indicate that no anomaly of any kind exists in the output of the model as no erratic behaviour was noticed when the simulation was performed.

Behaviour reproduction test

For the behaviour reproduction test, Sterman [18] suggests that the computation of some statistical measures includes descriptive statistics (e.g. R^2). Similarly, Barlas

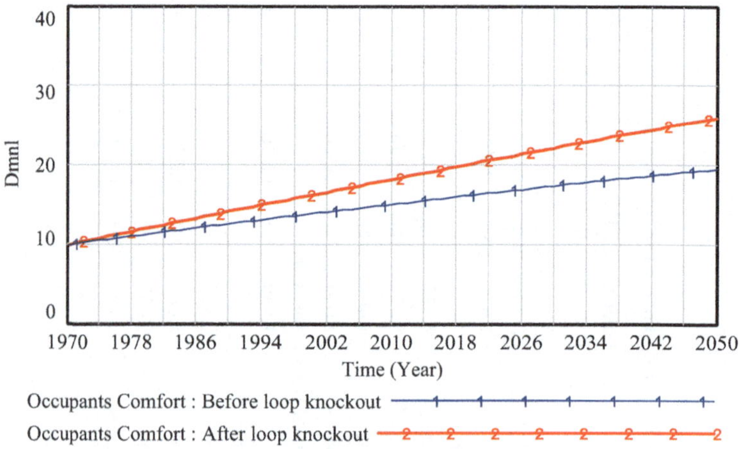

Fig. 6.33 Effect of loop knockout on occupants' comfort

Table 6.9 Change in household carbon emissions by end-use based on 'baseline' scenario for the year 2050 relative to 1990

	Total annual household energy consumption (KWh)						
	1990	2000	2010	2020	2030	2040	2050
'Baseline' scenario	491.2	546.8	519.4	445.3	413.1	385.4	360.6
Business-as-usual scenario of Johnston [11]	–	556.9	555.0	547.8	530.1	511.4	437.9

Table 6.10 Change in household carbon emissions by end-use based on 'baseline' scenario for the year 2050 relative to 1990

	Total annual household carbon emissions (million tonnes of CO_2)						
	1990	2000	2010	2020	2030	2040	2050
'Baseline' scenario	171.0	164.2	140.1	121.3	109.2	98.0	87.3
Business-as-usual scenario of Johnston [11]	–	132.7	132.5	137.1	127.7	118.0	97.5

[3] recommends trend comparison between the model output and actual (historical) data by formulating a linear, quadratic, or an exponential trend; comparing the periods by performing an autocorrelation function (ACF) test; and comparing the means by determining percentage error in the means.

In order to build confidence on the simulation output, the first behaviour reproduction test that was conducted is to compare the output of the 'baseline' scenario in this book with the results of Johnston's [11] 'business-as-usual' scenario. The Johnston's [11] 'business-as-usual' scenario is based on the current trends of energy efficiency improvements as at the time the research was conducted. Most of the assumptions made by Johnston [11] for the scenario are similar to this model's 'baseline' scenario assumptions. The results of this comparative analysis are summarised in Tables 6.9 and 6.10 for household energy consumption and household carbon emissions, respectively. The results shown in Tables 6.9 and 6.10 for the total annual household energy consumption and carbon emissions display the same pattern of trend. There are, however, some differences in the two models. The values of total annual household energy consumption are lower than those of Johnston [11].

This may possibly be due to the fact that the evidence from the historical data utilised by the developed model in this book suggests a drastic reduction in energy consumption. This is because of different government schemes regarding energy consumption yielding positive results. These differences are likely due to different assumptions made, input data utilised, or the modelling philosophy employed by the two models. Regarding the different assumptions made, the two models clearly show that there are some differences in household carbon emissions (see Table 6.10). This difference may be as a result of the assumption made by this model regarding the energy to carbon conversion factor. Also, the utilised input

Table 6.11 Total household carbon emissions

Time (year)	Historical (million tonnes)	Simulation (million tonnes)	Error (%)
1970	177.69	181.90	2.37
1975	171.66	169.11	−1.48
1980	167.77	167.34	−0.26
1985	164.17	166.16	1.21
1990	155.44	161.01	3.58
1995	145.29	157.95	8.72
2000	148.96	154.22	3.53
2005	154.59	161.90	4.73
2010	148.96	140.09	−5.96

data are different. Johnston [11] used data from a number of sources, basically from Shorrock and Dunster (1997) as published by BRE. This research book also utilised data from different sources, basically from Palmer and Cooper [14] as published by the DECC. It should be noted that the DECC is the government body housing energy data in the UK. Data used in this research book are more recent than those of Johnston [11]. This may therefore account for the differences. Finally, it may be due to the modelling philosophy used by the two models. Johnston's model was implemented using Excel template based on building physics, which only utilises quantitative data; while the model in this research book is implemented using the SD approach, which is based on the feedback control. It is worth mentioning that the SD modelling approach utilised in this research book combined both the quantitative and qualitative data sources; this is seen as the main strength of this approach.

Furthermore, the total carbon emissions were used as an example to see how the simulated results mimic the historical data. As such, the percentage error of the simulated results indicates a maximum of 8.72% error for 1995. Since the maximum percentage error is less than ±10%, this shows that the simulated results are consistent with the historical data (Table 6.11).

Additionally, it is equally important not only to identify the size of error, but also to identify its sources. Theil's inequality statistic is one method by means of which errors can be resolved into systematic and random portions. Theil's inequality statistic is derived from a decomposition of mean squared error (MSE). This statistic identifies the proportion of MSE that is attributable to bias ($\boldsymbol{U^m}$), unequal variance ($\boldsymbol{U^s}$), and unequal covariance ($\boldsymbol{U^c}$). Sterman [18] argued that a fit between the historically observed and simulated data can be compared to determine the size of error and its composition using root mean squared percentage error (RMSPE) and Theil's inequality. MSE, RMSPE, U^m, U^s, and U^c values were computed using Eqs. 6.11–6.15.

$$\text{MSE} = \frac{1}{n}\sum_{t=1}^{n}(S_t - H_t)_2 \tag{6.11}$$

$$\text{RMSPE} = \sqrt{\frac{1}{n}\sum_{t=1}^{n}\left[\frac{(S_t - H_t)}{H_t}\right]^2} \tag{6.12}$$

$$U^m = \frac{(\overline{S}_t - \overline{H}_t)^2}{\frac{1}{n}\sum_{t=1}^{n}(S_t - H_t)^2} \tag{6.13}$$

$$U^s = \frac{(s_s - s_h)^2}{\frac{1}{n}\sum_{t=1}^{n}(S_t - H_t)^2} \tag{6.14}$$

$$U^c = \frac{2(1 - r)s_s s_h}{\frac{1}{n}\sum_{t=1}^{n}(S_t - H_t)^2} \tag{6.15}$$

where n = number of observations ($t = 1,....n$); S_t = simulated value at time t; H_t = historical value at time t; $\overline{S}_t$ = mean simulated value at time t; $\overline{H}_t$ = mean historical value at time t; s_s = standard deviation of simulated value; s_h = standard deviation of historical value; and r = correlation coefficient between simulated and historical values.

Table 6.12 shows the summary statistics used in measuring the level of correspondence between the simulated and historical values of the total household carbon emissions. The results indicate that less than 1% of the mean squared error is attributable to bias. Only about 18% of the error results in unequal variation, leaving about 82% of the error attributable to unequal covariation. The implication of this is that there is very little systematic error in the results, which indicates that the simulated value tracks the historical trend near perfectly.

Furthermore, the trend analysis of some variables in the model was conducted using the autoregressive integrated moving averages (ARIMA) statistic. This statistic used the model output(s) of this research book and compared them to the actual (historical) data using two variables of interest. The variables selected are the

Table 6.12 Summary statistics measuring level of correspondence between the simulated and historical values of total household carbon emissions

$\sum_{t=1}^{n}(S_t - H_t)^2$	3.797×10^{14}	$\overline{S}_t$	162.19×10^6	r	0.851
$\sum_{t=1}^{n}\left[\frac{(S_t-H_t)}{H_t}\right]^2$	2.001×10^{-2}	$\overline{H}_t$	159.38×10^6	U^m	0.002
MSE	4.219×10^{13}	S_S	10.76×10^6	U^s	0.177
RMSPE	6.003×10^{-2}	S_h	10.80×10^6	U^c	0.821

Table 6.13 Validation based on statistical significance of behaviour

Model	Model fit statistics			Ljung–Box Q			Number of outliers
	R^2	MAPE	MaxAPE	Statistics	DF	*p* value	
Model-actual (average annual carbon emissions per household)	0.991	0.900	3.494	14.366	18	0.741	0
Model-actual (average annual energy consumption per household)	0.999	0.074	0.482	9.757	18	0.940	0

MAPE mean absolute percentage error, *MaxAPE* maximum absolute percentage error, *DF* degree of freedom, *p* value significance value

'average annual energy consumption per household' and 'average annual carbon emissions per household'. The choice of these variables is dictated by the fact that the main output of this research is both household energy consumption and household carbon emissions. However, the historical time series data available on carbon emissions are not disaggregated as was done for energy consumption end-uses. Therefore, the 'average annual energy consumption per household' and 'average annual carbon emissions per household' are chosen as the test variables. The 'goodness-of-fit' (*R* squared) results for the two variables of interest are shown in Table 6.13. The results suggest a good 'goodness of fit' for the two variables explored based on the values of R^2 as 0.991 and 0.999 for average annual carbon emissions per household and average annual energy consumption per household, respectively.

The results also show the mean absolute percentage error (MAPE) for the two variables as being very small (MAPE = 0.900% *for average annual carbon emissions per household, and* MAPE = 0.074% *for average annual energy consumption per household*) (Table 6.13).

Swanson et al. [20] suggest a MAPE of less than 10% as being very good. Based on this suggestion, it is then safe to say that the maximum absolute percentage error (MaxAPE) computed for the two variables as indicated in Table 6.13 is adequate as well (MaxAPE = 3.494% *for average annual* CO_2 *emissions per household, and* MaxAPE = 0.482% *for average annual energy consumption per household*).

Further to these tests, ACF tests were conducted in order to detect any significant errors in the periods of the time series for the model outputs and actual data by testing a hypothesis. If the sample ACF for simulated model is represented by $r_s(k)$ and the one for the actual (historical) data is $r_a(k)$, then the null hypothesis is formulated thus:

$$H_0 : r_s(1) - r_a(1) = 0, r_s(2) - r_a(2) = 0, \ldots, r_s(M) - r_a(M) = 0 \quad (6.16)$$

and the alternative hypothesis is:

$$H_1 : r_s(k) - r_a(k) \neq 0 \text{ for at least one } k \quad (6.17)$$

where k is any pair of values.

It must be noted that $[r_s(k) - r_a(k)] = 0$ under the H_0 and as such an interval is constructed based on the standard error (SE) of the difference of $r_s(k) - r_a(k)$ under which the H_0 is rejected should the difference $[r_s(k) - r_a(k)]$ fall outside the interval:

$$\begin{cases} -SE[r_s(k) - r_a(k)], \\ SE[r_s(k) - r_a(k)], \end{cases} \quad (6.18)$$

In order to take a decision regarding the hypothesis, the same decision rule as formulated under the Eqs. 6.1 and 6.2 was used. That is, the significance value (p value) is compared to the significance level ($\alpha = 0.05$), and the null hypothesis is either rejected or accepted. That is, if the p value is less than the significance level, the null hypothesis is rejected (if p value $< \alpha$, reject the null); otherwise the null hypothesis is accepted.

In this case, the Ljung–Box Q statistics computed for the two variables of interest as shown in Table 6.13 reveals that the p value for average annual carbon

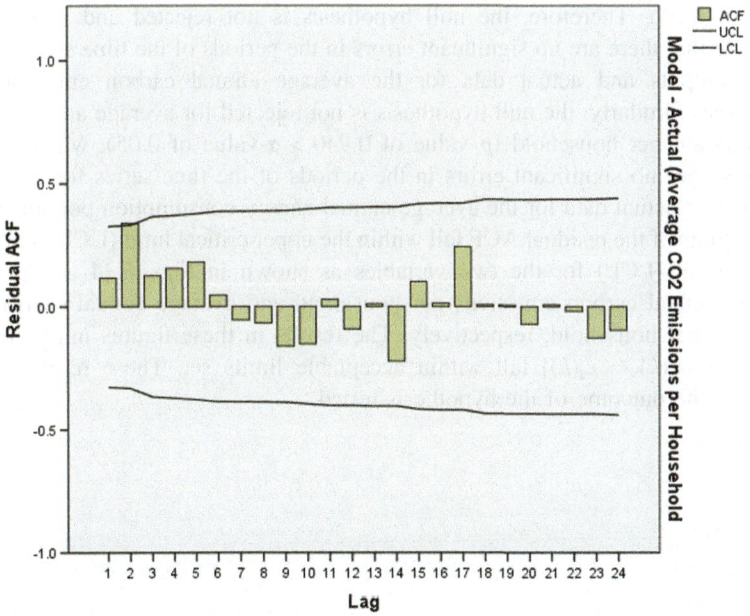

Fig. 6.34 Plots of residual autocorrelation function for carbon emissions

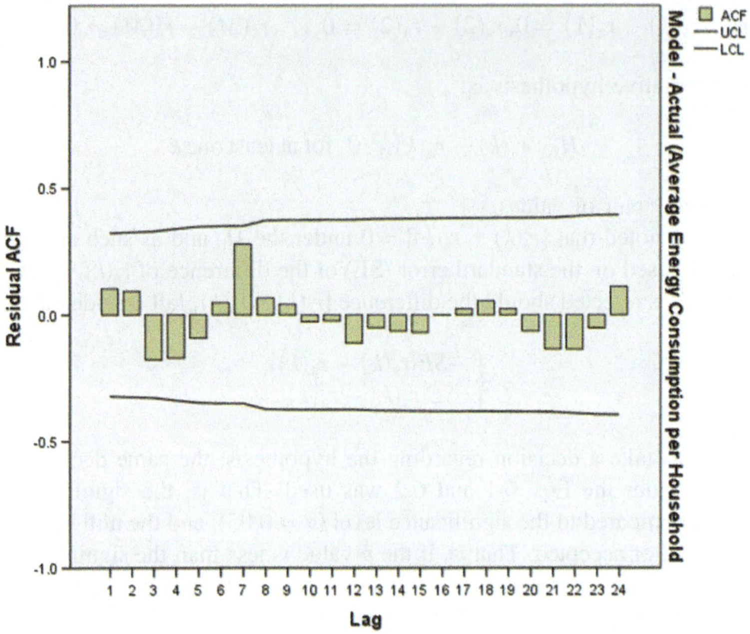

Fig. 6.35 Plots of residual autocorrelation function for energy consumption

emissions per household is greater than the significance level (p value of $0.741 > \alpha$-value of 0.05). Therefore, the null hypothesis is not rejected and it is safe to conclude that there are no significant errors in the periods of the time series for the model outputs and actual data for the average annual carbon emissions per household. Similarly, the null hypothesis is not rejected for average annual energy consumption per household (p value of $0.940 > \alpha$-value of 0.05), which implies that there are no significant errors in the periods of the time series for the model outputs and actual data for the average annual energy consumption per household.

The plots of the residual ACF fall within the upper critical limit (UCL) and lower critical limit (LCL) for the two variables as shown in Figs. 6.34 and 6.35 for average annual carbon emissions per household and average annual energy consumption per household, respectively. The results in these figures imply that the values of $[r_s(k) - r_a(k)]$ fall within acceptable limits set. These results further reinforce the outcome of the hypothesis tested.

6.7 Summary

This chapter has described and discussed the general assumptions made under the 'baseline' scenario. The model behaviour under this scenario suggested the way by which household energy consumption and carbon emissions attributable to the UK housing stock would evolve over the years under the assumptions that the current energy efficiency measures, consumption behaviour, and energy prices trends will be sustained. The results of key variables from each of the modules were discussed. In the population/household module, the model behaviour indicated that the total UK population is on an upward trajectory until 2050. Also, the number of households in the UK was predicted by the model as likely to grow on a yearly average of 1.02%, while the average household size tends towards two per household by the year 2050. Under the dwelling internal heat module, the model output suggests that both the dwelling internal heat and dwelling internal temperature will continue to grow. These are owing to improvements envisaged in dwellings' thermal performance, thereby increasing dwellings' airtightness and the desire to improve thermal comfort by raising the temperature set-point by householders.

Furthermore, the chapter discussed the insights observed from key variables under the occupants' thermal comfort module. These include the behaviour of the perceived dwelling temperature that the model output suggests will grow over the years until 2050. The reason for this trend is as offered under the discussion of dwelling internal heat and dwelling internal temperature. This is seen to result in improved occupants' thermal comfort. Within the climatic-economic-energy efficiency interaction module, the model results suggest that the unfavourable climatic effects will decline as a result of efforts aiming at reducing carbon emissions. However, the model result suggests that the energy efficiency standard improvement on dwelling energy efficiency will tend to increase. The results from the scenario suggest that reductions in household energy consumption of approximately 9 and 27% are visible by the years 2020 and 2050, respectively, below the year 1990 levels. These translate to savings of 29 and 49% in carbon emissions by the years 2020 and 2050, respectively. The insights from the model show that the greatest savings in both household energy consumption and carbon emissions are expected from space and water heating.

This chapter has also described and discussed the testing and validation exercises conducted for the model. The chapter discussed the validation done with 15 experts in the field of energy and SD. Further, the chapter described and discussed the tests performed within the SD modelling software, Vensim. These tests are categorised into the structure-oriented and behaviour pattern tests. For the structure-oriented tests, the following tests were performed: the boundary adequacy tests, structure assessment test, dimensional consistency test, parameter assessment test, extreme conditions test, and integration error test. Also, the behaviour pattern tests include the behaviour anomaly test, behaviour reproduction test, and behavioural sensitivity analysis. The results of all these tests yielded positive answers, confirming that the model satisfies all the rules and regulations of the SD approach and the model output(s) are consistent with the real system.

References

1. Banfill PFG, Jenkins DP, Patidar S, Gul M, Menzies GF, Gibson GJ (2012) The risk of buildings overheating in a low-carbon climate change future. In: Proceedings of international conference for enhanced building operations, University of Salford, Manchester, United Kingdom, 23–26 Oct
2. Barlas Y (1985) Validation of system dynamics models with a sequential procedure involving multiple quantitative methods. Unpublished Ph.D. dissertation, Georgia Institute of Technology, Atlanta, GA
3. Barlas Y (1996) Formal aspects of model validity and validation in system dynamics. Syst Dyn Rev 12(3):183–210
4. Building Research Establishment (2012) The government's standard assessment procedure for energy rating of dwellings. Garston, Watford
5. Chartered Institution of Building Services Engineers (2013) The limits of thermal comfort: avoiding overheating in European buildings (CIBSE TM52). CIBSE, London
6. Coyle RG (1977) Management system dynamics. Wiley, Chichester
7. Coyle RG (1997) System dynamics modelling: a practical approach. Chapman & Hall, UK
8. EST, DECC, DEFRA (2012) Powering the nation: household electricity-using habits revealed. EST/DECC/DEFRA, London
9. Forrester JW, Senge PM (1980) Tests for building confidence in system dynamics models. In: Legasto AA, Forrester JW, Lyneis JM (eds) System dynamics. North Holland, Amsterdam
10. Groesser SN, Schwaninger M (2012) Contribution to model validation: hierarchy, process, and cessation. Syst Dyn Rev 28(2):157–181
11. Johnston D (2003) A physically based energy and carbon dioxide emission model of the UK housing stock. Ph.D. thesis, Leeds Metropolitan University, UK
12. Motawa I, Oladokun M (2015) A model for the complexity of household energy. Energy Build 87:313–323
13. Oladokun MG, Odesola IA (2015) Modelling the complex sociotechnical systems of household energy and carbon emissions: a system dynamics approach. J Eng Technol 6 (1):121–143
14. Palmer J, Cooper I (2012) United Kingdom housing energy fact file. DECC, London
15. Ranganath BJ, Rodrigues LLR (2008) System dynamics: theory and case studies. I.K. International Publishing House, New Delhi
16. Richardson GP, Pugh A (1999) Introduction to system dynamics: modelling with DYNAMO. MIT Press, Cambridge
17. Rogelj J, Meinshausen M, Knutti R (2012) Global warming under old and new scenarios using IPCC climate sensitivity range estimates. Nat Climate Change 2(4):248–253
18. Sterman J (2000) Business dynamics: systems thinking and modelling for a complex world. Irwin McGraw-Hill, Boston
19. Summerfield AJ, Lowe RJ, Oreszczyn T (2010) Two models for benchmarking UK domestic delivered energy. Building Res Inf 38(1):12–24
20. Swanson DA, Tayman J, Bryan TM (2011) MAPE-R: a rescaled measure of accuracy for cross-sectional subnational population forecasts. J Population Res 22(2–3):225–243

Chapter 7
Policy Analysis

7.1 Introduction

In Chap. 6, the 'baseline' scenario was presented which communicates the most probable way in which the household energy consumption and carbon emissions of the UK housing stock will evolve over the years starting from 1970 until 2050. This is based on the assumption that the trends depicted by historical data will continue in that way. Evidently, if there is any policy change in the future that is clearly different from the current ones, the profile(s?) of household energy consumption and carbon emissions could be altered. This may result in producing a set of entirely different consumption–emissions profiles of the UK housing stock. For example, the UK Government may decide to implement a stringent energy efficiency policy if it is apparent that the current energy efficiency is unlikely to yield the required legally binding carbon emissions' reduction targets. As such, this chapter uses the discussions under the 'baseline' scenario to develop a few illustrative scenarios that may evolve as a result of some policy changes. Many scenarios can be assumed and for the purpose of illustration, four hypothetical scenarios are assumed that include 'efficiency', 'behavioural change', 'economic', and 'integrated' scenarios. In all of these scenarios, the future of household energy consumption and carbon emissions for the entire UK housing stock is explored. As such, the household energy consumption carbon emissions' reductions attributable to each of the scenarios are given and discussed. The rest of the chapter explains the underlying assumptions for each of the scenarios and discusses the results emanating from them.

M. G. Oladokun and C. O. Aigbavboa, *Simulation-Based Analysis of Energy and Carbon Emissions in the Housing Sector*, Green Energy and Technology,
https://doi.org/10.1007/978-3-319-75346-1_7

7.2 Policy Description and Assumptions

This section of the book describes some illustrative policy assumptions. Four different scenarios are attempted. The assumptions related to each of the scenario are discussed as follows.

7.2.1 *'Efficiency' Scenario Assumptions*

This scenario is based on the assumptions made under the 'baseline' scenario, but describes a situation whereby more evidence regarding the negative impact of climate change as a result of greenhouse emissions continues to emerge. This new evidence about the threat of the negative effect of climate change is assumed to continue. This will further reinforce the need to meet the legally binding carbon reduction targets as set by the government, for example reduction targets based on the Climate Change Act of 2008 in the UK. Also, the scenario assumes that more evidence will emerge on the possibility of not meeting the legally binding carbon reduction targets as the report of European Union (EU) entitled EU Study Predicts Clean Energy, Climate Failure by 2050 [1] suggests. Based on this new evidence, it then portends to trigger more stringent energy efficiency measures in order to cut carbon emissions significantly.

Energy efficiency measures are then assumed to concentrate on household dwellings, and this is technology led. As such, it is assumed that there will be improvements in the uptake of dwelling insulation measures, thereby resulting in each household's dwelling being thermally insulated with increasing energy efficiency ratings of dwellings such as the SAP rating. Further to these assumptions, the airtightness of dwellings will increase and it is again assumed that a great deal of energy savings will arise from this scenario. However, occupants will offset any savings that would have been made by seeking more thermal comfort, thereby increasing their dwellings' internal temperature set-point from 19 to 21 °C. At this stage, 'standard' consumption behaviour is still assumed to be maintained by the householders as done under the 'baseline' scenario. Fabric insulation depicted as 'insulation factor' in the model is therefore assumed to increase by 25% beyond the levels set under the 'baseline' scenario.

Additionally, a structural adjustment is made to the model as shown in Fig. 7.1. Under the 'baseline' scenario, the 'SAP rating' of the dwellings in the UK is qualitatively modelled within the SD and this was used to directly model the 'effect of energy efficiency standard improvement on dwelling energy efficiency'. However, in this scenario, it is assumed that the 'SAP rating' will gradually increase until 2040 at a gentle slope of 0.75% beyond the level it was in the year 2010. This is assumed not to go beyond 0.95 (in a scale of 0–1). To achieve this qualitatively, an intermediary dummy variable was introduced within the model as shown in Fig. 7.1. The model equation for this is also shown in Eq. 7.1.

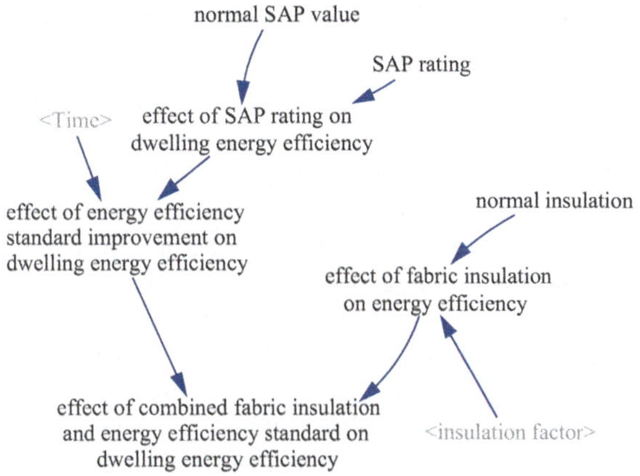

Fig. 7.1 Structural adjustment to the climatic-economic-energy efficiency interaction module

$$\begin{aligned}&\text{effect of energy efficiency standard improvement on dwelling}\\&\text{energy efficiency} = \text{IF THEN ELSE}(\text{Time} \le 2010, \text{effect of SAP}\\&\text{rating on energy efficiency standard improvement, MIN}\\&(0.95, \text{effect of SAP rating on energy efficiency standard}\\&\text{improvement} + \text{RAMP}(0.0075, 2010, 2040)))\end{aligned} \quad (7.1)$$

7.2.2 'Behavioural Change' Scenario Assumptions

This scenario is also based on major assumptions made under the 'baseline' scenario. The effects of occupants' behavioural change on household energy consumption and carbon emissions are the main policy drivers that this scenario illustrates. And as such, frugal consumption behaviour is emphasised by this scenario. That is, their daily habitual behaviours tend towards energy saving in their homes. In addition, it is assumed that this will have an effect on the dwelling internal temperature set-point as maintained by the occupants. A set-point of the dwelling internal temperature is therefore assumed to be 18.5 °C. Also within this scenario, energy prices are assumed to increase somewhat/marginally, thereby causing the energy bills paid by the householders to increase slightly by 5% beyond the level assumed under the 'baseline' scenario. In terms of energy efficiency, the assumption of this scenario is similar to that of the 'baseline' scenario, namely that no many substantive changes are made to the current trends in household energy efficiency apart from the continuation of the existing trends. For this scenario, no attempt is made to change any of the parameters within the population/household module. Therefore, no special effects are anticipated from the 'number of households' and 'average household size' other than their profile(s?) as generated internally within the model.

7.2.3 'Economic' Scenario Assumptions

The 'economic' scenario is also based on the 'baseline' scenario, but with the emphasis on the effects of energy bills on household energy consumption and carbon emissions. The scenario describes a future in which the UK Government will formulate a policy freezing the energy prices in order to score some political points. It is assumed this will cause a reduction in energy bills payable by the householders. The scenario anticipates the likelihood of this dip in energy bills to free up more disposable income for householders in an attempt to lower the number of those in fuel poverty. With this, the scenario assumes that the householders will seek more thermal comfort as a result of more disposable income, thereby increasing their dwelling internal temperature set-point slightly, though with 'standard' consumption behaviour. It is necessary to state that the scenario has the potential of illustrating the impact of energy prices' surge or dip on the household energy consumption and carbon emissions. Therefore, for this scenario, all other variables are kept as they were for the 'baseline' scenario apart from the following changes made to some of the parameters within the model. The '% increment in energy bills' is set at −5%, and the dwelling internal temperature set-point is set to 20 °C.

7.2.4 'Integrated' Scenario Assumptions

The 'integrated' scenario integrates and harmonises the assumptions made under the 'efficiency', 'behavioural change', and 'economic' scenarios as they impact on the household energy consumption and carbon emissions of the UK housing stock. The scenario assumes that the energy efficiency improvements as described and emphasised under the 'efficiency' scenario will be maintained. Further, the scenario assumes that householders will display frugal energy consumption behaviour. As such, they are interested in monitoring their energy usage at home. That is, they exercise some behavioural habits aimed at saving energy consumption at home such as turning down heating in vacant rooms, washing at a lower temperature, and the like. With all these, they are, however, assumed to maintain a dwelling internal temperature set-point of 20 °C. Additionally, within this period, energy prices are expected to be frozen as explained under the 'economic' scenario in Sect. 7.2.3.

7.3 Policy Analysis for Energy Consumption

In this section of the book, details about the outputs from the simulation run for different scenarios are presented and discussed.

7.3.1 *Analysis of the 'Efficiency' Scenario Results for Household Energy Consumption*

Figures 7.2, 7.3, 7.4, 7.5 and 7.6 capture the behaviour of energy consumption per household by end-uses as attributable to space heating, hot water, appliances, lighting, and cooking based on the'efficiency' scenario. Also, Figs. 7.7 and 7.8 show the graph of the respective average and total annual household energy consumption under the efficiency scenario. Visually, the graphs (Figs. 7.2, 7.3, 7.4, 7.5, 7.6, 7.7 and 7.8) display a downward trend for all the variables indicated in the graphs, thereby showing that there are reductions in household energy consumption across the board. However, the graphs show little information on the extent to which household energy consumption has been reduced in the years 2020 and 2050 relative to the year 1990. This information is necessary in order to know whether the reduction targets based on the Climate Change Act of 2008 are achieved by this scenario. To this end, more analysis of the simulation results is carried out to provide further insights into the reduction targets achieved.

The results of further analysis conducted on the household energy consumption based on end-uses for the entire UK housing stock is shown in Tables 7.1 and 7.2 illustrating the changes in household energy for the year 2020 and 2050 relative to the year 1990, respectively.

For the year 2020, the results of analysis of 'efficiency' scenario as shown in Table 7.1 suggest that the total household energy consumption for the entire UK housing stock is expected to be reduced by about 60.59 TWh of energy representing a 12.33% reduction relative to the year 1990 levels. Furthermore, the analysis suggests that apart from the household appliances energy that is expected to surge within this period by about 30%, others are expected to decline within this

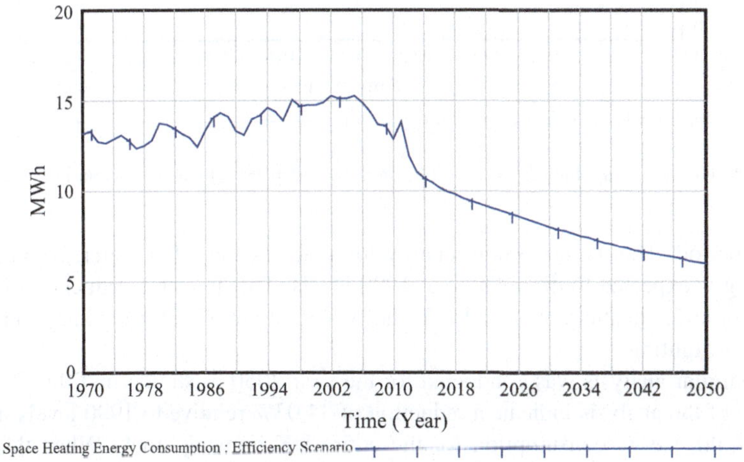

Fig. 7.2 Household space heating energy consumption under the 'efficiency' scenario

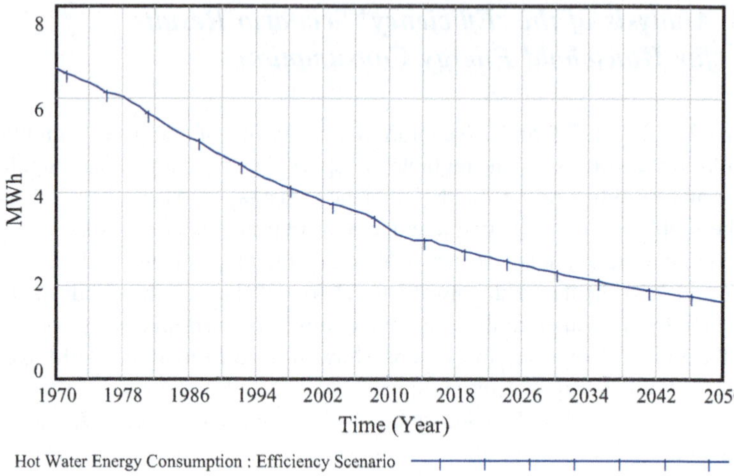

Fig. 7.3 Household water heating energy consumption under the 'efficiency' scenario

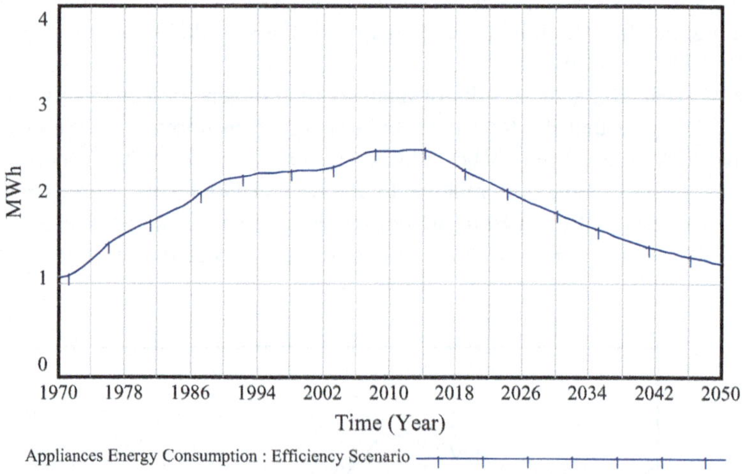

Fig. 7.4 Household appliances energy consumption under the 'efficiency' scenario

same period. That is, household energy for space heating, hot water, cooking and lighting is expected to dip relative to 1990 levels. This is with a reduction of about 11% for space heating, about 29% for hot water, about 47% for cooking and about 15% for lighting.

A similar analysis was conducted for the year 2050 as shown in Table 7.2. The results of the analysis indicate a reduction of 34.03% relative to 1990 levels in total household energy consumption for the entire UK housing stock. When this total household energy is disaggregated by end-use, the model results suggest that the

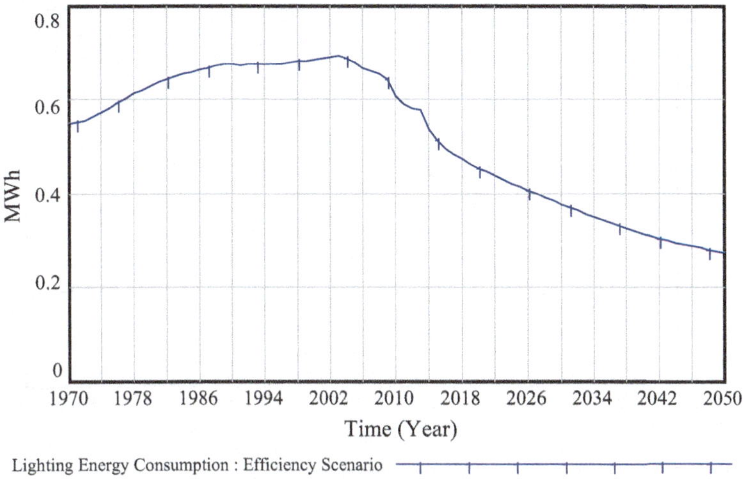

Fig. 7.5 Household lighting energy consumption under the 'efficiency' scenario

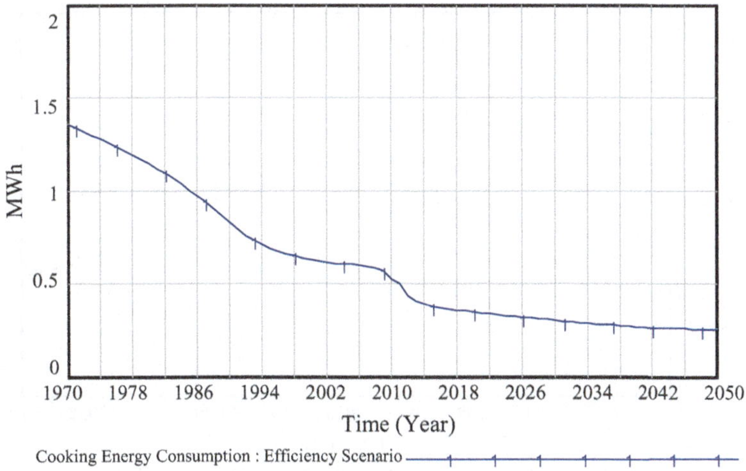

Fig. 7.6 Household cooking energy consumption under the 'efficiency' scenario

household energy consumption attributable to space heating is anticipated to be reduced by 94.89 TWh, which is a reduction of approximately 32% compared to 1990 levels. Furthermore, the household energy consumption due to hot water is expected to be reduced by about 50.32 TWh, which amounts to a reduction of approximately 47%, again relative to the base case 1990. Additionally, the model results suggest that cooking energy is expected to decline by about 54%, while the lighting energy is expected to witness a fall of about 39%. Paradoxically, household energy consumption for appliances declines by about 12% for the year 2050 relative

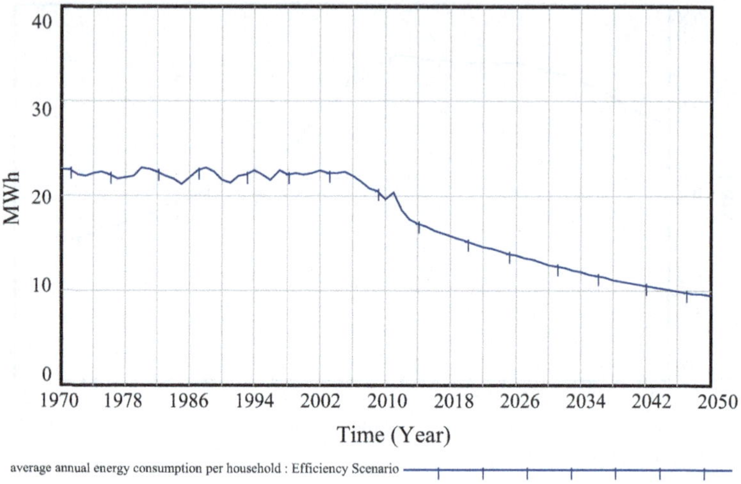

Fig. 7.7 Average annual household energy consumption under the 'efficiency' scenario

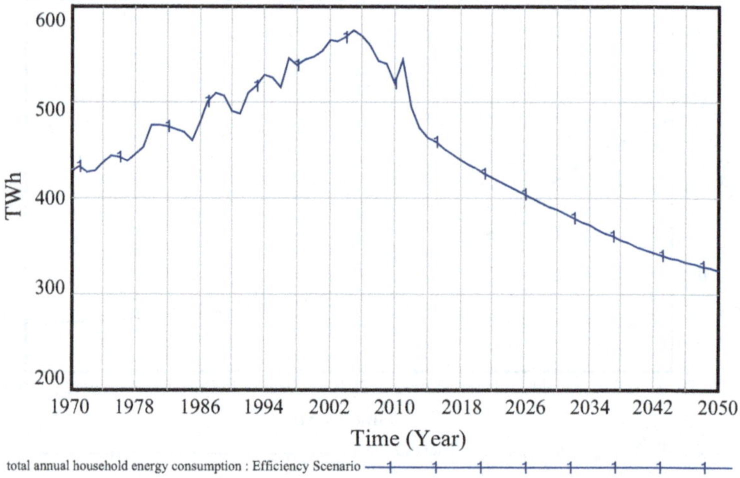

Fig. 7.8 Total annual household energy consumption under the 'efficiency' scenario

to 1990 levels, which increased/s? by about 30% for the year 2020 as discussed above. Ordinarily, this is expected to surge under the guise of technological improvements in household appliances. This is premised on the possibility of an increment in its adoption rate among the householders with an increasing number of households.

Changes to the parameters and structure of the model as relate to the fabric insulation and energy efficiency improvement for the 'efficiency' scenario are expected to reduce household energy consumption. The results in Tables 7.1 and

Table 7.1 Change in household energy consumption by end-uses based on 'efficiency' scenario for the year 2020 relative to 1990

	Household energy consumption (1990) (TWh)	Household energy consumption (2020) (TWh)	Change in household energy consumption (TWh)[a]	Percentage change in household energy consumption (%)[a]
Space heating	300.92	268.84	−32.08	−10.66
Hot water	108.20	76.74	−31.46	−29.08
Cooking	18.88	9.97	−8.91	−47.19
Lighting	15.29	12.96	−2.33	−15.24
Appliances	47.93	62.14	+14.21	+29.65
Total	491.22	430.63	−60.59	−12.33

[a]Relative to 1990 base as enshrined in Climate Change Act of 2008

Table 7.2 Change in household energy consumption by end-uses based on 'efficiency' scenario for the year 2050 relative to 1990

	Household energy consumption (1990) (TWh)	Household energy consumption (2050) (TWh)	Change in household energy consumption (TWh)[a]	Percentage change in household energy consumption (%)[a]
Space heating	300.92	206.03	−94.89	−31.53
Hot water	108.20	57.88	−50.32	−46.51
Cooking	18.88	8.70	−10.18	−53.92
Lighting	15.29	9.41	−5.88	−38.46
Appliances	47.93	42.06	−5.87	−12.25
Total	491.22	324.07	−167.15	−34.03

[a]Relative to 1990 base as enshrined in Climate Change Act of 2008

7.2 suggest that there are reductions in energy consumption apart from that of appliances for the year 2020. These percentage reductions are then compared with the 'baseline' scenario results as shown in Table 7.3. While the results generally show reductions in the total household energy and household energy by end-use, the amount of these reductions to meet the necessary carbon emissions reduction targets for the year 2020 and 2050 is therefore unlikely. This is due to the fact that efforts aiming at only the fabric insulation and energy efficiency improvement in general cannot bring about the required level of savings in energy consumption. In fact, a rebound effect is even likely to set in, in that the savings made may be expended on reaching an improved thermal comfort.

Table 7.3 Comparison of 'efficiency' scenario with 'baseline' scenario results for the percentage reductions in household energy consumption

	'Baseline' scenario (2020)	'Efficiency' scenario (2020)	'Baseline' scenario (2050)	'Efficiency' scenario (2050)
Space heating	−7.99	−10.66	−25.38	−31.53
Hot water	−26.61	−29.08	−41.10	−46.51
Cooking	−44.28	−47.19	−44.92	−53.92
Lighting	−10.92	−15.24	−25.18	−38.46
Appliances	+35.34	+29.65	+5.21	−12.25
Total	−9.35	−12.33	−26.60	−34.03

7.3.2 Analysis of the 'Behavioural Change' Scenario Results for Household Energy Consumption

The behaviour of household energy consumption based on end-uses is shown in Fig. 7.9. Also, the behaviour of average and total annual household energy consumption is depicted in Fig. 7.10. A visual inspection on the two figures indicates that household energy consumption under the 'behavioural change' scenario is identical to the household energy consumption under the 'efficiency' scenario illustrated in Figs. 7.2 and 7.8 above as the trends tend to decline as they approach 2050. However, they are different in terms of the level of reductions achieved by 2020 and 2050. The reductions in the 'behavioural change' scenario are more pronounced, and as such, the insights there portend that they are likely to be more than the reductions achieved under the 'efficiency' scenario.

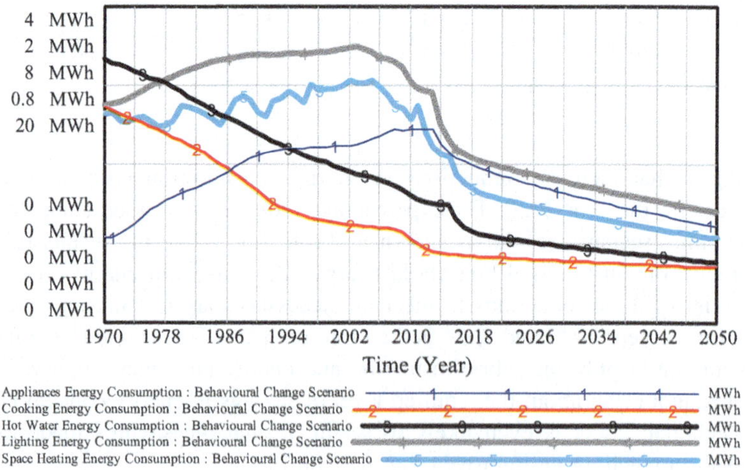

Fig. 7.9 Behaviour of household energy consumption based on end-uses under the 'behavioural change' scenario

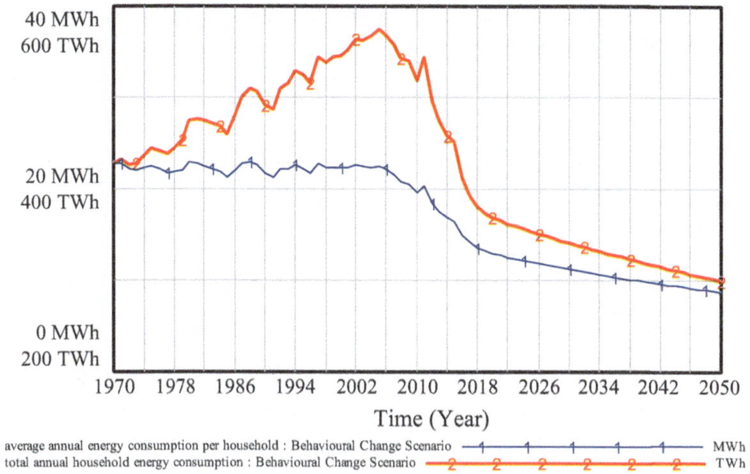

Fig. 7.10 Behaviour of average and total annual household energy consumption under the 'behavioural change' scenario

Tables 7.4 and 7.5 show the results of the further analysis carried out. As was done for the 'baseline' and 'efficiency' scenarios, the household energy consumption for the entire UK housing stock is analysed for the year 2020 and 2050 relative to the year 1990, respectively (Tables 7.4 and 7.5). The consumption profile is unchanged as such when compared to the 'efficiency' scenario because the space heating is still the one responsible for the largest portion with a reduction of about 72.20 and 117.01 TWh for 2020 and 2050, respectively. This is followed by hot water, which accounts for reductions of 45.85 and 55.94 TWh based on expectation for the years 2020 and 2050, respectively. For these two years, cooking and lighting energy are expected to decline as well, although the reductions anticipated from the results are minimal.

However, household energy consumption for appliances grew by 6.64 TWh beyond 1990 levels in 2020 as experienced in the 'efficiency' scenario. This is likely to be attributed to the fact that the number of households continues to increase, thereby increasing the number of household appliance acquisitions during this period. By 2050, however, the scenario results suggest that the appliances' energy consumption will be marginally reduced below 1990 levels (Table 7.5).

The profound insight from the analysis conducted reveals that the savings in the total household energy consumption under the 'behaviour change' scenario are generally more than those of 'efficiency' and 'baseline' scenarios. These results reinforce the comment of Janda [2] that '... *buildings don't use energy; people do*'. The result implies that occupants' behavioural change has the capability of contributing to the household energy consumption reductions and consequently contributes to carbon emissions' reduction targets of the government in conjunction with other policy frameworks.

Table 7.4 Change in household energy consumption by end-uses based on 'behavioural change' scenario for the year 2020 relative to 1990

	Household energy consumption (1990) (TWh)	Household energy consumption (2020) (TWh)	Change in household energy consumption (TWh)[a]	Percentage change in household energy consumption (%)[a]
Space heating	300.92	228.72	−72.20	−23.99
Hot water	108.20	62.35	−45.85	−42.38
Cooking	18.88	11.88	−7.00	−37.08
Lighting	15.29	11.44	−3.85	−25.18
Appliances	47.93	54.57	6.64	13.85
Total	491.22	368.95	−122.27	−24.89

[a]Relative to 1990 base as enshrined in Climate Change Act of 2008

Table 7.5 Change in household energy consumption by end-uses based on 'behavioural change' scenario for the year 2050 relative to 1990

	Household energy consumption (1990) (TWh)	Household energy consumption (2050) (TWh)	Change in household energy consumption (TWh)[a]	Percentage change in household energy consumption (%)[a]
Space heating	300.92	183.91	−117.01	−38.88
Hot water	108.20	52.26	−55.94	−51.70
Cooking	18.88	12.06	−6.82	−36.12
Lighting	15.29	9.53	−5.76	−37.67
Appliances	47.93	42.07	−5.86	−12.23
Total	491.22	299.84	−191.38	−38.96

[a]Relative to 1990 base as enshrined in Climate Change Act of 2008

7.3.3 Analysis of the 'Economic' Scenario Results for Household Energy Consumption

The behaviour exhibited by the household energy consumption in terms of end-uses and average and total annual household energy consumption are illustrated in Figs. 7.11 and 7.12, respectively. Similarly, to other scenarios, a visual inspection of the two figures shows that the trends of household energy consumption under the 'economic' scenario are on a downward trajectory as they tend towards 2050. The slopes are gentler as the graphs tend towards 2050 when compared with both the 'efficiency' and 'behavioural change' scenarios. The reductions in 'economic' scenario seem subtle and the insights there indicate/anticipate that they are unlikely

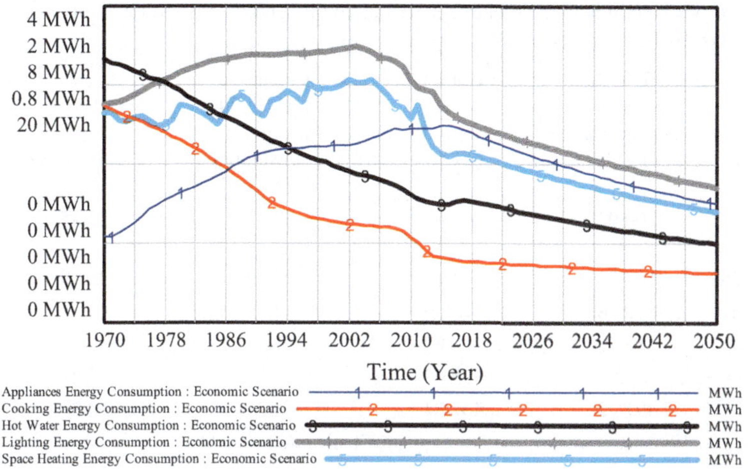

Fig. 7.11 Behaviour of household energy consumption based on end-uses under the 'economic' scenario

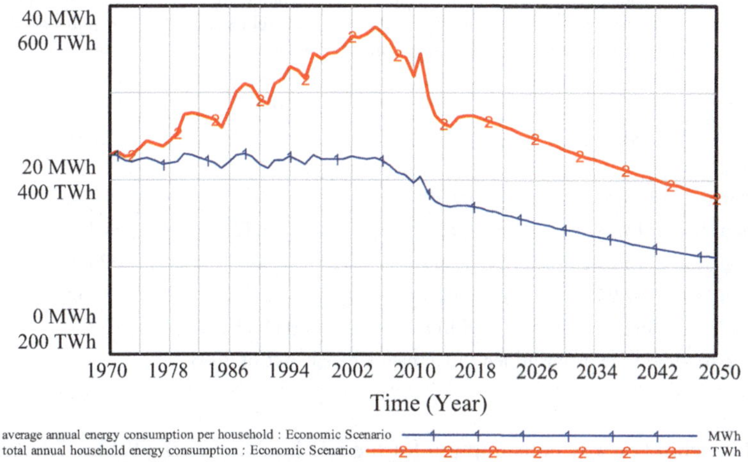

Fig. 7.12 Behaviour of average and total annual household energy consumption under the 'economic' scenario

to meet the required reductions target. Therefore, a detailed analysis is required to reveal any hidden insights.

Tables 7.6 and 7.7 show the results of a detailed analysis conducted on household energy consumption of the UK housing stock for the years 2020 and 2050, respectively, as done for other previous scenarios. Interestingly, the results of simulation for the scenario show that reductions of only approximately 10.87 and 22.62% are expected to be taken place in the years 2020 and 2050, respectively, in

Table 7.6 Change in household energy consumption by end-uses based on 'economic' scenario for the year 2020 relative to 1990

	Household energy consumption (1990) (TWh)	Household energy consumption (2020) (TWh)	Change in household energy consumption (TWh)[a]	Percentage change in household energy consumption (%)[a]
Space heating	300.92	263.45	−37.47	−12.45
Hot water	108.20	84.19	−24.01	−22.19
Cooking	18.88	10.69	−8.19	−43.38
Lighting	15.29	13.82	−1.47	−9.61
Appliances	47.93	65.69	17.76	37.05
Total	491.22	467.84	−53.38	−10.87

[a]Relative to 1990 base as enshrined in Climate Change Act of 2008

Table 7.7 Change in household energy consumption by end-uses based on 'economic' scenario for the year 2050 relative to 1990

	Household energy consumption (1990) (TWh)	Household energy consumption (2050) (TWh)	Change in household energy consumption (TWh)[a]	Percentage change in household energy consumption (%)[a]
Space heating	300.92	238.75	−62.17	−20.66
Hot water	108.20	67.79	−40.41	−37.35
Cooking	18.88	10.58	−8.30	−43.96
Lighting	15.29	11.66	−3.63	−23.74
Appliances	47.93	51.33	3.40	7.09
Total	491.22	380.11	−111.11	−22.62

[a]Relative to 1990 base as enshrined in Climate Change Act of 2008

total household energy consumption for the UK housing stock (Tables 7.6 and 7.7). These results indicate reduced savings when compared to the results of the 'efficiency' and 'behavioural change' scenarios in Sects. 7.3.1 and 7.3.2. The results are therefore not surprising as they are expected because of the assumptions of the scenario relating to energy prices reduction. A reduction in energy bills frees up more disposable income to householders, which they will not mind spending on improved thermal comfort at home. This allows occupants to raise their dwellings' internal temperature set-point. Also, freezing energy prices by the government threaten energy security and innovative investments in clean energy by energy service providers. All these explain the insights from the results.

Additionally, the consumption profile remains unchanged as such as witnessed in the previous simulated scenarios. This is mainly due to the fact that space heating still accounts for the largest amount of savings in household energy consumption. The savings here are approximately 37.47 and 62.17 TWh for 2020 and 2050, respectively. This is closely followed by hot water with reductions of 24.01 and 40.41 TWh as expected for the years 2020 and 2050, respectively. Both the cooking and lighting energy are expected to decline as well. As usual, household energy consumption for appliances is expected to grow by about 17.76 TWh beyond 1990 levels in 2020 (Table 7.6). Again, this increase may be attributed to a rise in the number of households, thereby increasing the amount of household appliance acquisition during this period. In addition, the availability of more disposable income as a result of the reduction in energy bills may encourage householders to purchase more home appliances, which invariably results in increased energy consumption. By 2050, however, the scenario results suggest that the appliances energy consumption will drop marginally to below the 1990 levels (Table 7.7). Again, there are no suitable past studies with which to compare the output of this scenario.

7.3.4 Analysis of the 'Integrated' Scenario Results for Household Energy Consumption

Figures 7.13 and 7.14 display the behaviours of household energy consumption based on end-uses, and the behaviours of average and total annual household energy consumption per household in the UK housing stock. By conducting a quick visual inspection of Fig. 7.13, it can be observed that the trends are similar to those displayed by the 'behavioural change' scenario. This indicates that the household energy consumption under the 'integrated' scenario tends to decline as 2050 is approached. Similarly, the profile of average and total annual household energy consumption as shown in Fig. 7.14 also tends to decline towards 2050.

Energy savings from these graphs when compared to that of the 'behavioural change' scenario are almost the same, as few or no differences are noticed by the visual inspection. However, there is a well-pronounced difference between the behaviour of this scenario (in terms of household energy) when compared to corresponding figures under the 'efficiency' scenario. The insights from the scenario then show that there is a possibility of more savings from this scenario than from the 'efficiency' scenario. But, it is difficult to make this kind of conjecture when compared to the 'behavioural change' scenario.

As for all the previous scenarios, a further analysis is carried out in order to bring to light all the hidden insights from the scenario that Figs. 7.13 and 7.14 cannot explain. The household energy consumption for the UK housing stock is analysed for the years 2020 and 2050 relative to the year 1990 and is shown in Tables 7.8 and 7.9, respectively. The profile of the contribution of each household energy end-use to energy savings for this scenario remains unchanged when compared to the previous scenarios because the space heating is still responsible for the major

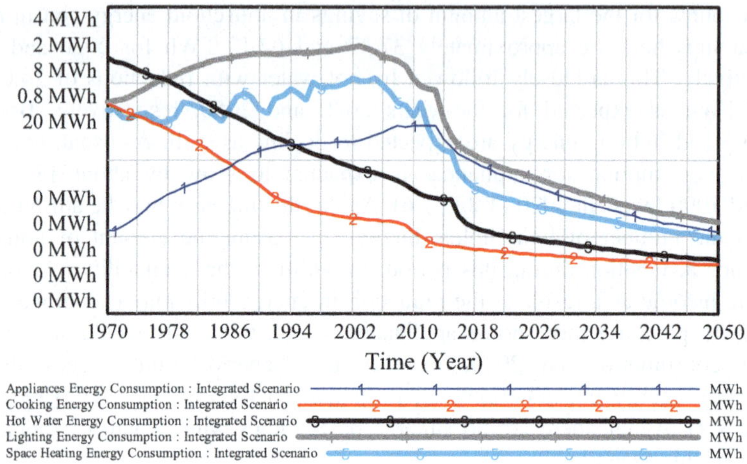

Fig. 7.13 Behaviour of household energy consumption based on end-uses under the 'integrated' scenario

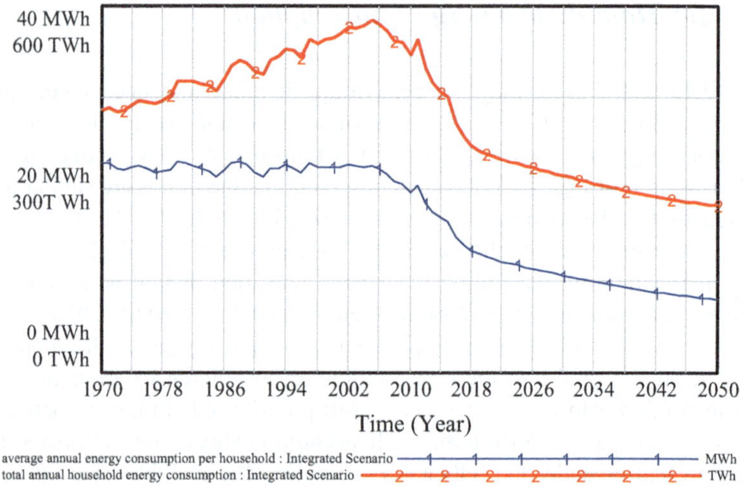

Fig. 7.14 Behaviour of average and total annual household energy consumption under the 'integrated' scenario

share of savings expected in household energy consumption. For example, the results of this scenario suggest that reductions of approximately 79.31 and 131.56 TWh are anticipated by the years 2020 and 2050, respectively. As expected, this is followed by hot water with a savings of 48.38 and 60.48 TWh for the years 2020 and 2050, respectively (Tables 7.8 and 7.9, respectively). These figures show that they are lower than the savings anticipated under both the 'efficiency' and 'behavioural change' scenarios.

Table 7.8 Change in household energy consumption by end-uses based on 'integrated' scenario for the year 2020 relative to 1990

	Household energy consumption (1990) (TWh)	Household energy consumption (2020) (TWh)	Change in household energy consumption (TWh)[a]	Percentage change in household energy consumption (%)[a]
Space heating	300.92	221.61	−79.31	−26.36
Hot water	108.20	59.82	−48.38	−44.71
Cooking	18.88	11.66	−7.22	−38.24
Lighting	15.29	11.18	−4.11	−26.88
Appliances	47.93	53.49	5.56	11.60
Total	491.22	357.76	−133.46	−27.17

[a]Relative to 1990 base as enshrined in Climate Change Act of 2008

Table 7.9 Change in household energy consumption by end-uses based on 'integrated' scenario for the year 2050 relative to 1990

	Household energy consumption (1990) (TWh)	Household energy consumption (2050) (TWh)	Change in household energy consumption (TWh)[a]	Percentage change in household energy consumption (%)[a]
Space heating	300.92	169.36	−131.56	−43.72
Hot water	108.20	47.72	−60.48	−55.90
Cooking	18.88	10.93	−7.95	−42.11
Lighting	15.29	8.18	−7.11	−46.50
Appliances	47.93	36.52	−11.41	−23.81
Total	491.22	272.70	−218.52	−44.49

[a]Relative to 1990 base as enshrined in Climate Change Act of 2008

Correspondingly, cooking and lighting energy is also expected to follow the trend of space and water heating. However, the volume of the savings anticipated is small as only 7.22 and 4.11 TWh are expected to be saved in cooking and lighting energy consumption, respectively, for the year 2020. The results of household energy consumption for appliances are not surprising when compared to the previous trends. For the year 2020, no savings are recorded below the levels of the year 1990. Again, the likely reason for this insight is the fact that the number of households continues to increase (as the model results suggest), thereby increasing the number of household appliances' acquisition and consequently more energy is used. However, the scenario results suggest that the appliances' energy consumption will be reduced to below 1990 levels by 11.41 TWh by the year 2050 (Table 7.9).

The analysis of results is illuminating as Table 7.8 shows that a total savings of about 27% below 1990 levels are expected by the year 2020, whereas a total savings of about 44% is anticipated by 2050. These figures are higher than those of any of the previous scenarios.

7.4 Policy Analysis for Carbon Emissions

7.4.1 Analysis of the 'Efficiency' Scenario Results for Household Carbon Emissions

Household carbon emissions by end-uses, and total and average annual household carbon emissions are shown in Figs. 7.15 and 7.16. respectively. The behaviour of efficiency scenario simulation results for household carbon emissions based on end-uses (Fig. 7.15) is similar to those shown for household energy consumption by end-uses (Figs. 7.2, 7.3, 7.4, 7.5 and 7.6). The graph shows a downward trend for carbon emissions for all the end-uses. Therefore, this indicates reductions in the carbon emissions' profile for space heating, hot water, cooking, lighting, and appliances.

Similarly, the insights from the average and total annual household carbon emissions (Fig. 7.16) follow a downward trend. These trends correspond to the same trends witnessed in average and total annual household energy consumption as shown in Figs. 7.7 and 7.8. The similarity is based on the reason advanced in Chap. 6 of this book, namely that there is a strong correlation between energy consumption and carbon emissions. However, the behaviour of space heating

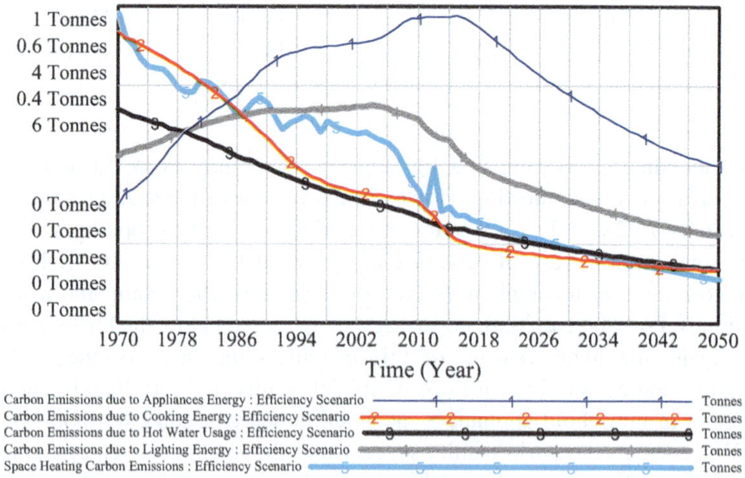

Fig. 7.15 Household carbon emissions by end-uses under the 'efficiency' scenario

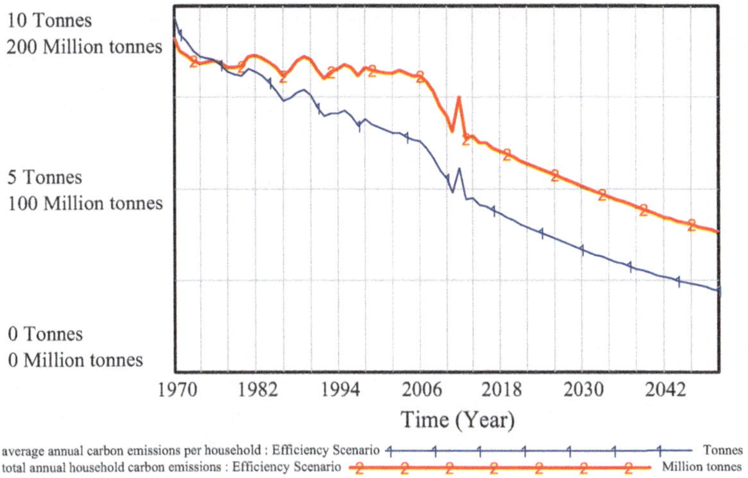

Fig. 7.16 Total and average annual household carbon emissions under the 'efficiency' scenario

carbon emissions plays a central role in explaining the behaviour displayed by both the average annual and total annual carbon emissions for the UK housing stock. This is in consonance with the observation of Palmer and Cooper [3] as well.

As expected for the average annual carbon emissions per household and total annual household carbon emissions, they have been moving in a downward direction since 1970. The two have fallen significantly since 1970, and the model suggests that the trend is anticipated to be sustained until 2050, based on the carbon reductions' agenda of the government. However, Fig. 7.16 shows little detail about the amount of reductions achieved under this scenario. Hence, a further analysis is necessary.

The amount of reductions in carbon emissions anticipated in the years 2020 and 2050 are, respectively, shown in Tables 7.10 and 7.11. The results suggest that the largest amount of reductions is expected to come from space heating, namely 42.79 and 65.71 million tonnes of CO_2 for 2020 and 2050, respectively. The results also indicate that a substantial amount of reduction is expected from hot water as well. Both space and water heating therefore remain the dominant end-uses where much of the reduction to meet the carbon emissions' targets is anticipated. Correspondingly, some reductions are also anticipated in cooking and lighting below the 1990 levels for the years 2020 and 2050 as, respectively, shown in Tables 7.10 and 7.11. However, for appliances, the expectation is a mixed one as the results show that no reductions in carbon emissions below the 1990 levels are anticipated. Although there are technological improvements in home appliances in terms of energy efficiency, this advancement could not be immediately translated

Table 7.10 Change in household carbon emissions by end-uses based on 'efficiency' scenario for the year 2020 relative to 1990

	Household carbon emissions (1990) (million tonnes of CO_2)	Household carbon emissions (2020) (million tonnes of CO_2)	Change in household carbon emissions (million tonnes of CO_2)[a]	Percentage change in carbon emissions (%)[a]
Space heating	94.47	51.68	−42.79	−45.29
Hot water	44.15	31.15	−13.00	−29.44
Cooking	7.93	4.03	−3.90	−49.18
Lighting	6.04	5.28	−0.76	−12.58
Appliances	18.43	25.39	+6.96	37.76
Total	171.01	117.53	−53.48	−31.27

[a]Relative to 1990 base as enshrined in Climate Change Act of 2008

Table 7.11 Change in household carbon emissions by end-uses based on 'efficiency' scenario for the year 2050 relative to 1990

	Household carbon emissions (1990) (million tonnes of CO_2)	Household carbon emissions (2050) (million tonnes of CO_2)	Change in household carbon emissions (million tonnes of CO_2)[a]	Percentage change in carbon emissions (%)[a]
Space heating	94.47	28.76	−65.71	−69.56
Hot water	44.15	23.36	−20.79	−47.09
Cooking	7.93	3.47	−4.46	−56.24
Lighting	6.04	3.79	−2.25	−37.25
Appliances	18.43	17.00	−1.43	−7.76
Total	171.01	76.37	−94.64	−55.34

[a]Relative to 1990 base as enshrined in Climate Change Act of 2008

into much savings. However, by 2050 (Table 7.10), some savings are expected. Generally, it is clear from the results of this scenario that the carbon emissions' target for 2020 and 2050 will not be met.

The results show that savings of about 31 and 55% will be met for 2020 and 2050 as against minimum targets of 34 and 80%, respectively. The implication of this is that placing much of the emphasis on energy efficiency improvements alone without corresponding efforts on other aspects of policy target is unlikely to yield the required level of savings.

7.4.2 *Analysis of the 'Behavioural Change' Scenario Results for Household Carbon Emissions*

As before, Figs. 7.17 and 7.18 illustrate the profile of household carbon emissions by end-uses, and total and average annual household carbon emissions, respectively. The two figures display similar consumption behaviour as shown for the household energy consumption above. Again, the graphs show a downward trend for carbon emissions for all the end-uses till 2050 under a varying degree of reductions. This, therefore, indicates that a reduction in the carbon emissions' profile of the UK housing stock is anticipated.

Tables 7.12 and 7.13 clearly illustrate the amount of reduction in carbon emissions anticipated in the years 2020 and 2050, respectively. Similar results are observed when this scenario is compared to that of the 'efficiency' scenario. As for the 'efficiency' scenario, the largest portion of reductions is expected from space heating (50.71 and 70.12 million tonnes of CO_2 for 2020 and 2050, respectively). This is followed by the carbon emissions attributable to hot water (18.51 and 23.12 million tonnes of CO_2 for 2020 and 2050, respectively). It is therefore clear that both space and water heating still remain the dominant end-uses where the greatest reductions in carbon emissions targets are expected. Accordingly, some reductions are also anticipated in cooking and lighting below the 1990 levels for the years 2020 and 2050, similar to what the results for the 'efficiency' scenario suggest. Like the 'efficiency' scenario, the appliances energy consumption is expected to show no reductions in carbon emissions below the 1990 levels. The implication of this is that the government policy is required to address this trend.

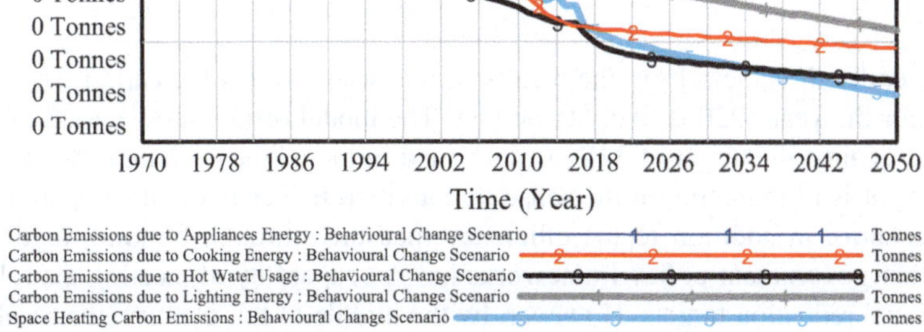

Fig. 7.17 Behaviour of household carbon emissions based on end-uses under the 'behavioural change' scenario

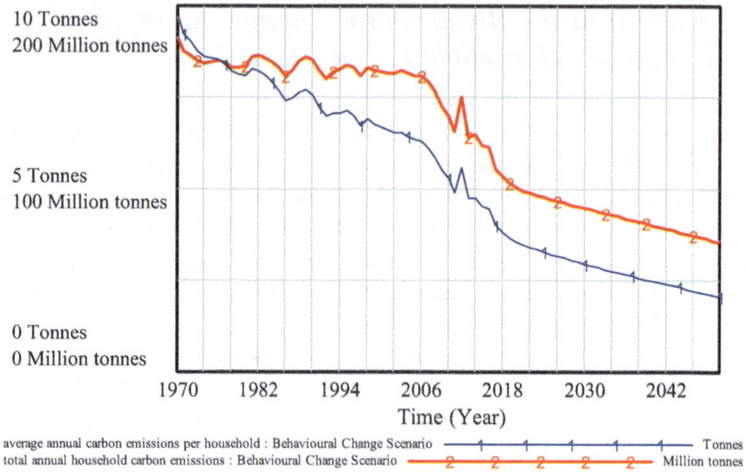

Fig. 7.18 Behaviour of average and total annual household carbon emissions under the 'behavioural change' scenario

Table 7.12 Change in household carbon emissions by end-uses based on 'behavioural change' scenario for the year 2020 relative to 1990

	Household carbon emissions (1990) (million tonnes of CO_2)	Household carbon emissions (2020) (million tonnes of CO_2)	Change in household carbon emissions (million tonnes of CO_2)[a]	Percentage change in carbon emissions (%)[a]
Space heating	94.47	43.76	−50.71	−53.68
Hot water	44.15	25.64	−18.51	−41.93
Cooking	7.93	4.75	−3.18	−40.10
Lighting	6.04	4.64	−1.40	−23.18
Appliances	18.43	22.19	3.76	20.40
Total	171.01	100.98	−70.03	−40.95

[a]Relative to 1990 base as enshrined in Climate Change Act of 2008

Generally, it is clear from the results of this scenario that the carbon emissions' target for the year 2020 is likely to be met. The model results show a total of about 41% as against the target of 34% as enshrined in the Climate Change Act of 2008. This result is illuminating in the sense that a vigorous behavioural campaign, as in this scenario, in addition to the efficiency measures through building regulations and other government policy frameworks has the capability of meeting the relevant emissions' reduction targets. However, by the middle of this century, the results for this scenario indicate it is unlikely to meet the carbon emissions' reduction target of 80%. The results show that only about 58% of the carbon emissions reductions are

Table 7.13 Change in household carbon emissions by end-uses based on 'behavioural change' scenario for the year 2050 relative to 1990

	Household carbon emissions (1990) (million tonnes of CO_2)	Household carbon emissions (2050) (million tonnes of CO_2)	Change in household carbon emissions (million tonnes of CO_2)[a]	Percentage change in carbon emissions (%)[a]
Space heating	94.47	24.35	−70.12	−74.22
Hot water	44.15	21.03	−23.12	−52.37
Cooking	7.93	4.81	−3.12	−39.34
Lighting	6.04	3.84	−2.20	−36.42
Appliances	18.43	16.99	−1.44	−7.81
Total	171.01	71.02	−99.99	−58.47

[a]Relative to 1990 base as enshrined in Climate Change Act of 2008

likely to be met. As suggested under the 'efficiency' scenario, the government policy should target other policy areas in addition to the 'behavioural change' in order to meet the required level of reductions.

It is important that the results of this scenario are not compared with model results from other studies because of the unavailability of the relevant previous results with which to compare.

7.4.3 *Analysis of the 'Economic' Scenario Results for Household Carbon Emissions*

The behaviour exhibited by the household carbon emissions by end-uses, and the total and average annual household carbon emissions are illustrated in Figs. 7.19 and 7.20, respectively. These results are again profound as they depict similar trends as demonstrated by those for household energy consumption. The emissions' profile displayed in the two figures indicates that towards the year 2050, the household carbon emissions are expected to fall. However, the rate of decline of the graphs is gentle, differentiating it somewhat from those of the 'efficiency' and 'behavioural change' scenarios. Based on the authors' conjecture, this gentle slope suggests that it is unlikely that this scenario will meet the legally binding carbon emissions' reduction targets of the government.

Tables 7.14 and 7.15 show the total amount of carbon emissions reductions envisaged by the scenario in the years 2020 and 2050, respectively. The results for this scenario suggest that the largest portion of reductions is expected to come from space heating in the amount of 38.04 and 59.18 million tonnes of CO_2 for 2020 and 2050, respectively. Also, it is evident from the results of this scenario that a substantial amount of reductions is expected from hot water as well in the amount of

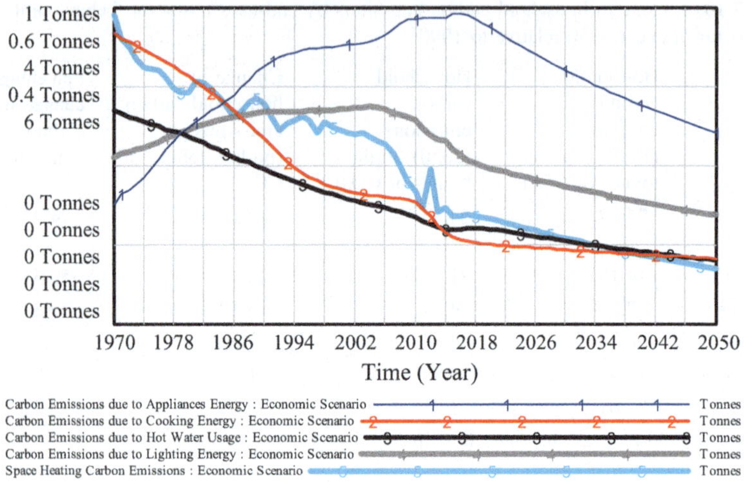

Fig. 7.19 Behaviour of household carbon emissions based on end-uses under the 'economic' scenario

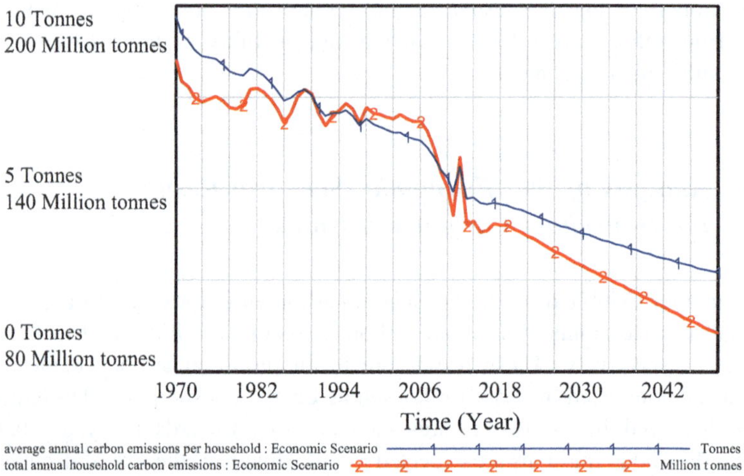

Fig. 7.20 Behaviour of average and total annual household carbon emissions under the 'economic' scenario

10.27 and 16.80 million tonnes of CO_2 for 2020 and 2050, respectively. These results reinforce the fact that both space and water heating remain the main end-uses where much of the reductions to meet the carbon emissions targets are anticipated as in other pervious scenarios. In the same way, savings in carbon emissions due to cooking and lighting are expected for the years 2020 and 2050, but the savings are limited when compared to both the space heating and hot water. Conversely, no

Table 7.14 Change in household carbon emissions by end-use based on 'economic' scenario for the year 2020 relative to 1990

	Household carbon emissions (1990) (million tonnes of CO_2)	Household carbon emissions (2020) (million tonnes of CO_2)	Change in household carbon emissions (million tonnes of CO_2)[a]	Percentage change in carbon emissions (%)[a]
Space heating	94.47	56.43	−38.04	−40.27
Hot water	44.15	33.88	−10.27	−23.26
Cooking	7.93	4.28	−3.65	−46.03
Lighting	6.04	5.58	−0.46	−7.62
Appliances	18.43	26.61	8.18	44.38
Total	171.01	126.78	−44.23	−25.86

[a]Relative to 1990 base as enshrined in Climate Change Act of 2008

Table 7.15 Change in household carbon emissions by end-use based on 'economic' scenario for the year 2050 relative to 1990

	Household carbon emissions (1990) (million tonnes of CO_2)	Household carbon emissions (2050) (million tonnes of CO_2)	Change in household carbon emissions (million tonnes of CO_2)[a]	Percentage change in carbon emissions (%)[a]
Space heating	94.47	35.29	−59.18	−62.64
Hot water	44.15	27.35	−16.80	−38.05
Cooking	7.93	4.23	−3.70	−46.66
Lighting	6.04	4.69	−1.35	−22.35
Appliances	18.43	20.71	2.28	12.37
Total	171.01	92.27	−78.74	−46.04

[a]Relative to 1990 base as enshrined in Climate Change Act of 2008

savings are expected from the carbon emissions due to appliances as the level emissions for this end-use are not anticipated to fall below the 1990 levels for both the years 2020 and 2050 as the results suggest in Tables 7.14 and 7.15, respectively. When the total carbon emissions' reductions for the years 2020 and 2050 are examined, it is evident from Tables 7.14 and 7.15 that the results of this scenario are unlikely to meet the carbon emissions' reduction targets as enshrined in the Climate Change Act of 2008. The results show that savings of approximately 25.86 and 46.04% are anticipated by the years 2020 and 2050 as against the minimum targets of 34 and 80% respectively. The implication of these results is profound since freezing energy prices as government policy in order to score some political points is to the detriment of meeting the required legally binding carbon emissions' reduction targets.

7.4.4 Analysis of the 'Integrated' Scenario Results for Household Carbon Emissions

As shown for all other previous scenarios, Figs. 7.21 and 7.22 illustrate the behaviour exhibited by the household carbon emissions by end-uses, and the total and average annual household carbon emissions, respectively. Again, both figures display a downward trend for carbon emissions for all the end-uses and the total and average annual household carbon emissions as they approach 2050 with a varying degree of savings. These downward trends show an ease sign of likelihood of achieving reductions in carbon emissions' profile of the UK housing stock. A detail analysis of these savings is carried out below.

It is clear from Tables 7.16 and 7.17 that the amount of reductions in household carbon emissions as expected in the years 2020 and 2050, respectively. The results are similar to those observed under the previous scenarios as the greatest reductions are expected from space heating (52.05 and 73.04 million tonnes of CO_2 for 2020 and 2050, respectively). As is to be expected, hot water contributes the second largest reductions in carbon emissions attributable to this end-use (19.41 and 24.97 million tonnes of CO_2 for 2020 and 2050, respectively). It is again apparent that both space and water heating are the dominant end-uses where the largest portion of reductions in carbon emissions targets is anticipated. As discussed under the household energy consumption, some reductions are also expected in cooking and lighting for the years 2020 and 2050, below the 1990 levels. These reductions are similar to what other previous scenarios suggest.

However, the model results for this scenario are similar to previous scenarios, namely that carbon emissions due to the appliances are unlikely to fall below the

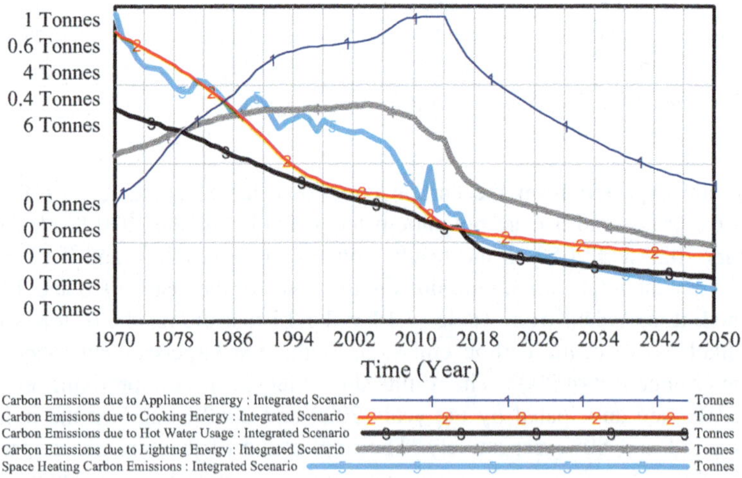

Fig. 7.21 Behaviour of household carbon emissions based on end-uses under the 'integrated' scenario

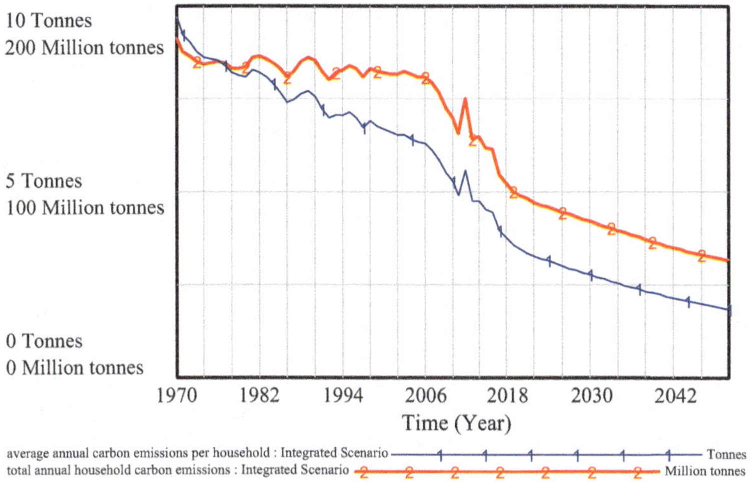

Fig. 7.22 Behaviour of average and total annual household carbon emissions under the 'integrated' scenario

Table 7.16 Change in household carbon emissions by end-use based on 'integrated' scenario for the year 2020 relative to 1990

	Household carbon emissions (1990) (million tonnes of CO_2)	Household carbon emissions (2020) (million tonnes of CO_2)	Change in household carbon emissions (million tonnes of CO_2)[a]	Percentage change in carbon emissions (%)[a]
Space heating	94.47	42.42	−52.05	−55.10
Hot water	44.15	24.74	−19.41	−43.96
Cooking	7.93	4.69	−3.24	−40.86
Lighting	6.04	4.57	−1.47	−24.34
Appliances	18.43	21.91	3.48	18.88
Total	171.01	98.33	−72.68	−42.50

[a]Relative to 1990 base as enshrined in Climate Change Act of 2008

1990 levels. Government policy is therefore required to overturn this trend with a deep cut in carbon emissions attributable to home appliances. However, the model results for this scenario in Table 7.17 suggest that the carbon emissions attributable to appliances are expected to fall below 1990 levels, thereby showing some sign of savings.

Generally, it is apparent from the results of this scenario that the carbon emissions' reduction target for the year 2020 is likely to be met. This is similar to the 'behavioural change' scenario. The result in Table 7.16 suggests that a total of approximately 43% of carbon emissions is expected as against the target of 34%

Table 7.17 Change in household carbon emissions by end-use based on 'integrated' scenario for the year 2050 relative to 1990

	Household carbon emissions (1990) (million tonnes of CO_2)	Household carbon emissions (2050) (million tonnes of CO_2)	Change in household carbon emissions (million tonnes of CO_2)[a]	Percentage change in carbon emissions (%)[a]
Space heating	94.47	21.43	−73.04	−77.32
Hot water	44.15	19.18	−24.97	−56.56
Cooking	7.93	4.36	−3.57	−45.02
Lighting	6.04	3.29	−2.75	−45.53
Appliances	18.43	14.75	−3.68	−19.97
Total	171.01	63.01	−108.00	−63.15

[a]Relative to 1990 base as enshrined in Climate Change Act of 2008

that is legally binding. This result therefore shows the efficacy of a vigorous behavioural change campaign, which in addition to the improvements in energy efficiency measures through stringent building regulations and other UK Government's policy frameworks, displays the capability of meeting the legally binding emissions' reduction targets.

Despite these giant strides shown by the scenario for the year 2020, it unlikely that the carbon emissions' reduction targets of 80% will be met by the middle of this century. The result indicates that carbon emissions' reductions of only approximately 63% are likely to be achieved. As suggested under the 'efficiency' and 'behavioural change' scenarios, the government policy needs to allow the market forces to dictate energy prices as this will ensure energy security and levels of commitment from the energy service provider to further invest in clean energy. Although the effect of growth in the number of households and decline in average household size is not independently explored by this model, the model has the capability of exploring these effects. However, government policy is required to moderate the growth in the number of households and the decline in the average household size as these two control the profile of the UK housing stock carbon emissions' profile.

7.4.5 Summary of the Results for All the Scenarios

The summary of the results for all the scenarios for household energy and carbon emissions is shown in Tables 7.18 and 7.19, respectively.

One of the most significant pieces of evidence from all the scenarios is that it is unlikely for any of the scenarios to individually meet the required legally binding reduction of an 80% cut in carbon emissions.

Table 7.18 Percentage reductions in household energy consumption for all scenarios

	'Baseline' scenario (%)	'Efficiency' scenario (%)	'Behavioural change' scenario (%)	'Economic' scenario (%)	'Integrated' scenario (%)
2020					
Space heating	−7.99	−10.66	−23.99	−12.45	−26.36
Hot water	−26.61	−29.08	−42.38	−22.19	−44.71
Cooking	−44.28	−47.19	−37.08	−43.38	−38.24
Lighting	−10.92	−15.24	−25.18	−9.61	−26.88
Appliances	+35.34	+29.65	13.85	37.05	11.60
Total	−9.35	−12.33	−24.89	−10.87	−27.17
2050					
Space heating	−25.38	−31.53	−38.88	−20.66	−43.72
Hot water	−41.10	−46.51	−51.70	−37.35	−55.90
Cooking	−44.92	−53.92	−36.12	−43.96	−42.11
Lighting	−25.18	−38.46	−37.67	−23.74	−46.50
Appliances	+5.21	−12.25	−12.23	7.09	−23.81
Total	−26.60	−34.03	−38.96	−22.62	−44.49

Table 7.19 Percentage reductions in household carbon emissions for all scenarios

	'Baseline' scenario (%)	'Efficiency' scenario (%)	'Behavioural change' scenario (%)	'Economic' scenario (%)	'Integrated' scenario (%)
2020					
Space heating	−43.70	−45.29	−53.68	−40.27	−55.10
Hot water	−27.32	−29.44	−41.93	−23.26	−43.96
Cooking	−46.91	−49.18	−40.10	−46.03	−40.86
Lighting	−8.94	−12.58	−23.18	−7.62	−24.34
Appliances	+42.65	37.76	20.40	44.38	18.88
Total	−29.08	−31.27	−40.95	−25.86	−42.50
2050					
Space heating	−65.64	−69.56	−74.22	−62.64	−77.32
Hot water	−41.77	−47.09	−52.37	−38.05	−56.56
Cooking	−47.54	−56.24	−39.34	−46.66	−45.02
Lighting	−23.68	−37.25	−36.42	−22.35	−45.53
Appliances	+10.42	−7.76	−7.81	12.37	−19.97
Total	−48.96	−55.34	−58.47	−46.04	−63.15

7.5 Summary

This chapter has described and extensively discussed some policy scenarios formulated in order to illustrate the use of the model in this research book. The illustrative scenarios demonstrated other ways by means of which household energy consumption and carbon emissions attributable to the UK housing stock would evolve over the years under different assumptions. The 'efficiency' scenario generally considers the effects of improvements in energy efficiency measures on household energy consumption and ultimately on household carbon emissions. Also, the 'behavioural change' scenario tries to model the effects of occupants' change of energy consumption behaviour on household energy consumption and carbon emissions profile. Further, the 'economic' scenario assumes a case of policy change by government favouring energy prices reduction, thereby reducing the energy bills payable by the householders and its consequences on household energy consumption and carbon emissions. And lastly, an 'integrated' scenario combines the assumptions in the first three scenarios and then analyses its effects on household energy consumption and carbon emissions.

The results from the 'efficiency' scenario suggest that reductions in household energy consumption of approximately 12 and 34% are likely to be visible by the years 2020 and 2050, respectively, below the year 1990 levels. The results further suggest that the household carbon emissions are likely to be reduced by about 31 and 55% in the years 2020 and 2050, respectively, below the 1990 levels. Additionally, the results of the 'behavioural change' scenario indicate that reductions in household energy consumption of approximately 25 and 39% are possible in the years 2020 and 2050, respectively, below the year 1990 levels. The model suggests that these reductions in household energy consumption are likely to translate to reductions in household carbon emissions of approximately 41 and 58% in the years 2020 and 2050, respectively, when compared to 1990 as the base case. Furthermore, the results for the 'economic' scenario show that reductions in household energy consumption of approximately 11 and 23% in the years 2020 and 2050, respectively, are possible when compared to the base year 1990. These translate to savings in carbon emissions of 26 and 46% by the years 2020 and 2050, respectively. Finally, the results from the 'integrated' scenario suggest that reductions in household energy consumption of approximately 27 and 44% are possible for the years 2020 and 2050, respectively. These figures amount to savings in household carbon emissions of approximately 43 and 63% for the years 2020 and 2050, respectively.

One of the most significant pieces of evidence from the scenarios is that the 34% carbon emissions target under the 'behavioural change' and 'integrated' scenarios is likely to be met under the vigorous campaign of energy consumption behavioural change of householders. However, the results for all the scenarios indicate that it is unlikely for any of the scenarios to meet the required legally binding reduction of an

80% cut in carbon emissions. Comparison of the model results with other studies clearly indicates there are some differences in the results. These are attributed to different assumptions made, different input data, and possibly the modelling philosophy employed.

References

1. European Union (2013) EU study predicts clean energy, climate failure by 2050. http://www.euractiv.com/energy/eu-study-predicts-climate-failur-news-532630, accessed on 16 Jan 2014
2. Janda KB (2011) Buildings don't use energy: people do. Archit Sci Rev 54:15–22
3. Palmer J, Cooper I (2012) United Kingdom housing energy fact file. DECC, London

Chapter 8
Conclusions and Reflections

8.1 Introduction

The aim of the research reported in this book is to develop a system dynamics-based model of the sociotechnical systems of energy consumption and carbon emissions with a focus on the UK housing sector. This chapter brings together the preceding chapters and the main conclusions that can be drawn from them to reflect the achievement of the purpose and objectives of this research book. Prior to highlighting the research originality and contribution to knowledge, the chapter provides details of the implications of research findings for research, practice, and society. The chapter also presents the limitations of the research in this book, and the recommendations for future research directions conclude the chapter.

8.2 Value for Practice

The developed model as the output of the research in this book has a number of implications for research, practice, and society.

Firstly, the study explored the complex intrinsic interrelationships among the STS of dwellings, occupants, and environment, as related to energy consumption and carbon emissions, by capturing their causes and effects. This is with the aim of improving the understanding of the complex system of household energy consumption and carbon emissions from a systems thinking perspective, thereby extending the knowledge base of system dynamics to household energy and carbon emissions. For example, the causal diagrams can lead to theory building by interested researchers. Additionally, the output of this research has the capability to spur research activities as mentioned under the recommendations for future research directions in Sect. 8.5.

M. G. Oladokun and C. O. Aigbavboa, *Simulation-Based Analysis of Energy and Carbon Emissions in the Housing Sector*, Green Energy and Technology,
https://doi.org/10.1007/978-3-319-75346-1_8

Secondly, the research in this book has implications for practice. The developed model in this research builds on the existing modelling efforts, which are traditionally restricted to building physics and regression-based forecasting, in order to generate new insights into the future using a nondeterministic systems approach. This then adds to the array of tools available in the field for practitioners. Since the developed model is highly transparent as all the variables and algorithms developed can be scrutinised, it has capacity to benefit the software developers significantly by prototyping it into other suitable user-friendly platforms.

Thirdly, the outcome of this research has implications for society by providing the policy makers with a decision-making tool upon which different scenarios regarding HECE can be tested before implementation.

8.3 Originality and Impacts

The originality of the research reported in this research book lies in the application of a system dynamics approach to capture and solve the complex problem relating to the future profiles of household energy consumption and carbon emissions by providing a policy advice tool to the policy makers. This is in an attempt to meet the carbon emission reduction targets as enshrined in the Climate Change Act of 2008. Consequently, the research effort within this book has made a number of contributions to knowledge. The unique contribution is the development of a model that incorporates sociotechnical issues that can be used for decision-making over time. Other contributions are highlighted below:

- The research in this book indicates the first system dynamics modelling efforts applied to the entire UK housing sector in order to model the household energy consumption and carbon emissions based on the complex sociotechnical interactions of the influencing variables.
- The model, especially the developed conceptual model—system thinking aspect, is capable of improving the theoretical knowledge base regarding the complex intrinsic interrelationships that exist among the sociotechnical influences of household energy consumption and carbon emissions.
- The developed model within this research book has the capability of being used to simulate and predict the future profiles of UK household energy consumption and carbon emissions over different time horizons.
- The developed model has the capability of producing a clear understanding of household energy consumption and the carbon emissions associated with it. This can serve as a decision-making policy tool with the capability to direct policy decisions by testing the effect different policy scenarios such as energy efficiency improvements and behavioural change are likely to have on household energy consumption and carbon emissions. The insights generated will allow policy makers to make informed decisions regarding any future policy

formulations concerning energy and carbon emissions within the UK housing sector.

- The developed model is also capable of modelling and exploring the potential rate at which the carbon emissions' reduction targets are being achieved within the UK housing sector.

8.4 Limitations

Any research effort relating to systems modelling, as is the case in this research book, is likely to have a number of strengths as well as suffering some drawbacks. The strengths of this research in terms of contribution to knowledge are stated in Sect. 8.3. The limitations of this research upon which further developments can be made are therefore summarised below:

- The research within this book is based purely on modelling the household energy consumption and carbon emissions (demand side). As such, the energy generation (energy supply side) attributable to the housing sector has not been considered or modelled by the model developed in this book. Therefore, the developed model cannot explicitly explore the technical improvements to the energy supply side in terms of the reduced carbon content of supplied energy as envisaged by these technical improvements. This then serves as a limitation to the research.
- The developed model in this book is an aggregated model of the entire UK households. As such, no attempt is made to disaggregate the household energy consumption and carbon emissions based on different dwelling types. Consequently, it is difficult to explore the energy consumption and carbon emissions' profiles attributable to the households within these different dwelling types. This is another limitation of this research. However, its inclusion will in no way affect the final results of the developed model.
- Another limitation of the research is in the energy-to-carbon conversion factor used in the developed model. It is believed that different fuels are used for different end-uses with different carbon emission factors. For example, householders are likely to use gas, electricity, or oil as fuel in order to achieve space and water heating in their dwellings. The carbon emission contents of each of these fuels differ. This research therefore used the same aggregated energy-to-carbon emission factors for all the household energy consumption end-uses. Changes to these carbon emissions factors will greatly alter the profile of household carbon emissions estimated by the developed model.
- The developed model in this research is limited in its application to other countries because it is specifically developed for the UK housing sector. This is mainly because the model algorithms are based on the UK data. However, other countries can benefit from it by domesticating the model algorithms.

8.5 Future Research Directions

The work within this book serves as a foundational system dynamics model upon which further research can be conducted in order to fully explore other options required to reduce the carbon emissions of the UK households. To this extent, the following areas of further research are recommended to be vigorously pursued in the coming years:

- Research on the energy supply side of the housing sector based on the system dynamics approach is encouraged. This is considered important because of the need to highlight the effect of clean energy supply, due to technological advancement, on the household energy consumption and carbon emissions. The output of such research can be linked to the developed model in this research to form an improved sociotechnical model of energy and carbon emissions of the UK housing sector.
- The current version of the developed model in this book is an aggregated model of the entire UK households. Another line of research can be followed by disaggregating this based on different dwelling types such as detached and semi-detached dwellings. This is in order to evaluate the carbon emissions' profiles attributable to each of these dwelling types. This line of research is considered important mainly because of the need to expand the capability and scope of the analysis performed by the developed model.
- The research within this book can be replicated with an expanded model boundary to accommodate all the variables that have been excluded by this current research. This will include the incorporation of the carbon emission factors of different fuels used for different end-uses of household energy consumption. Also, further research is considered important in order to improve the accuracy of the developed model.
- Additionally, it needs to be highlighted that the developed model within this book is specifically developed for the UK. As such, this research can be replicated in other developed countries. This is considered necessary in order for other developed countries to benefit from it by domesticating the model. Domesticating the model will entail using the data collected from the country replicating the model. This will be at the system conceptualisation and formulation stages of model development.

8.6 Summary

This chapter serves as the reflective summary of the research presented within this book as it brings together the findings of the research. The chapter has demonstrated how the aim and objectives of the research have been achieved and has given the main conclusions that follow the findings. The implications of the research for research, practice, and society were highlighted and discussed. Undoubtedly, the research effort within this book has made some contributions to knowledge, and these were accordingly highlighted in the chapter. Finally, the snapshots of the research limitations and recommendations for further research were given.

Appendix A
Model Algorithms

```
"% increament on energy bills"=
		-5
	Units: per cent [0,100,25]

appliances' energy:=
	GET XLS DATA ( 'Datasets.xlsx' , 'Datasets' , 'A' , 'M2' )
Units: MWh

Appliances' Energy Consumption= INTEG (
	(rate of appliances energy usage-energy to carbon conversion a),
		initial appliances usage)
Units: MWh

appliances' energy consumption for UK housing stock=
	Appliances Energy Consumption*households/10^6
Units: TWh

appliances' usage demand=
	2
Units: Dmnl

area of opening=
	10
Units: m*m
Area of windows and glazed doors

Artificial heat transfer=
	("discrepancy in int & setpoint temp"*insulation factor)+"total dwelling heat gains (dhg)"
/Time
Units: Watts/Year
```


M. G. Oladokun and C. O. Aigbavboa, *Simulation-Based Analysis of Energy and Carbon Emissions in the Housing Sector*, Green Energy and Technology,
https://doi.org/10.1007/978-3-319-75346-1

average annual carbon emissions per household=
Carbon Emissions due to Cooking Energy+Carbon Emissions due to Hot Water Usage
+Carbon Emissions due to Lighting Energy+Carbon Emissions due to Appliances Energy
+Space Heating Carbon Emissions
Units: Tonnes

average annual electricity bill:=
GET XLS DATA ('Datasets.xlsx' , 'Datasets' , 'A' , 'P2')
Units: £

average annual energy bills=
IF THEN ELSE(Time<=2010, ((average annual electricity bill+average annual gas bill
)+(5139.16-2.455*Time+164.363*weighted average energy prices+12.612
*effect of dwelling energy efficiency on energy bills))/2, FORECAST(((average annual electricity bill
+average annual gas bill)+(5139.16-2.455*Time+164.363*weighted average energy prices
+12.612
*effect of dwelling energy efficiency on energy bills))/2,39,-45))
Units: £

average annual energy consumption per household=
Cooking Energy Consumption+Hot Water Energy Consumption+Lighting Energy Consumption
+Space Heating Energy Consumption+Appliances Energy Consumption
Units: MWh

average annual gas bill:=
GET XLS DATA ('Datasets.xlsx' , 'Datasets' , 'A' , 'O2')
Units: £

average effect of solar gains=
0.9*area of opening*frame factor*average solar access factor*solar flux*solar transmittance factor for glazing
Units: Watts

average household size=
Population/households
Units: People/Households

average life expectancy=
78.8
Units: Year

average solar access factor=
0.7
Units: Dmnl [0.3,1,0.01]

average total fertility rate=
3
Units: Dmnl

births=

IF THEN ELSE(Time=population equilibrium time, deaths, IF THEN ELSE(Time<=2011
, average total fertility rate*Population
*0.08/reproductive time, FORECAST(average total fertility rate*Population
*0.08/reproductive time,39,100)))
Units: People/Year

carbon depletion=
Carbon Emissions due to Appliances Energy*carbon depletion factor
Units: Tonnes/Year

carbon depletion factor=
1.325
Units: 1/Year [-2,5,0.005]

Carbon Emissions due to Appliances Energy= INTEG (
(energy to carbon conversion -carbon depletion)*0.5246,
initial appliances usage carbon emissions)
Units: Tonnes

carbon emissions due to appliances energy of UK housing stock=
Carbon Emissions due to Appliances Energy*households/10^6
Units: Million tonnes

Carbon Emissions due to Cooking Energy= INTEG (
(energy to carbon conversion - carbon depletion)*0.5246,
initial cooking carbon emissions)
Units: Tonnes

carbon emissions due to cooking energy of UK housing stock=
Carbon Emissions due to Cooking Energy*households/10^6
Units: Million tonnes

carbon emissions due to hot water energy of UK housing stock=
Carbon Emissions due to Hot Water Usage*households/10^6
Units: Million tonnes

Carbon Emissions due to Hot Water Usage= INTEG (
(energy to carbon conversion hw-hw carbon depletion)*0.5246,
initial hot water usage carbon emissions)

Units: Tonnes

Carbon Emissions due to Lighting Energy= INTEG (
(energy to carbon conversion l-l carbon depletion)*0.5246,
initial lighting usage carbon emissions)
Units: Tonnes

carbon emissions due to lighting energy of UK housing stock=
Carbon Emissions due to Lighting Energy*households/10^6
Units: Million tonnes

carbon emissions due to space heating energy of UK housing stock=
Space Heating Carbon Emissions*households/10^6
Units: Million tonnes

cooking energy:=
GET XLS DATA ('Datasets.xlsx' , 'Datasets' , 'A' , 'K2')
Units: MWh

Cooking Energy Consumption= INTEG (
(cooking energy rate-energy to carbon conversion),
initial cooking energy)
Units: MWh

cooking energy consumption for UK housing stock=
Cooking Energy Consumption*households/10^6
Units: TWh

cooking energy rate=cooking energy*effect of energy efficiency standard on cooking energy/
effect of energy bills on energy consumption/0.88
-0.35*FORECAST(cooking energy/1.88, 40, e)) ,

Units: MWh/Year

deaths=
Population*mortality
Units: People/Year

demand for cooking energy=
2
Units: Dmnl

dhg due to appliances less cooking=
((207.8*(total floor area*average household size)*EXP(0.4714))*(1+0.157*COS (2*pi*(Time-1.178)))*1000/60)/5000
Units: Watts

dhg due to artificial lighting=
(59.73*(total floor area*average household size)^0.4714)*0.96^2*(1+0.5*COS (2*pi*(Time-0.2)))*0.85*1000/(24*30*12)
Units: Watts

dhg due to cooking=
35+(7*average household size)
Units: Watts

dhg due to no of people=
60*average household size
Units: Watts

dhg due to water heating=
80.5*average household size
Units: Watts

"discrepancy in int & ext temp"=
external air temp-dwelling int temp
Units: Deg Cent

"discrepancy in int & setpoint temp"=
IF THEN ELSE("% increament on energy bills"=0, setpoint temp-dwelling int temp
, IF THEN ELSE("% increament on energy bills"=25, setpoint temp-0.5-dwelling int temp
, IF THEN ELSE("% increament on energy bills"=50, setpoint temp-1-dwelling int temp
, IF THEN ELSE("% increament on energy bills"=75, setpoint temp-1.5-dwelling int temp
, setpoint temp-2-dwelling int temp))))
Units: Deg Cent

dwelling int temp=
Dwelling Internal Heat/(65*temperature conversion factor)
Units: Deg Cent

Dwelling Internal Heat= INTEG (
(Artificial heat transfer+Natural heat transfer),
130000)
Units: Watts

dwelling internal temp=
dwelling int temp*1.7
Units: Deg Cent

effect of combined fabric insulation and energy efficiency standard on dwelling energy efficiency
=
IF THEN ELSE(Time<2011, 1+(effect of energy efficiency standard improvement on dwelling energy efficiency
+effect of fabric insulation on energy efficiency
), 1+(effect of energy efficiency standard improvement on dwelling energy efficiency
+effect of fabric insulation on energy efficiency
)+RAMP(0.01,2010,2050))
Units: Dmnl

effect of dwelling energy efficiency on energy bills=
1/effect of combined fabric insulation and energy efficiency standard on dwelling energy efficiency
Units: Dmnl

effect of energy bills on energy consumption=
1-(1/((1+(1/average annual energy bills)*100)+effect of dwelling energy efficiency on energy bills
)-0.9)*(1-(1/(1+"% increament on energy bills"
/normal energy bills)))/effect of unfavourable climatic effects on international energy prices
Units: Dmnl

effect of energy efficiency on hot water energy=
1/effect of combined fabric insulation and energy efficiency standard on dwelling energy efficiency
Units: Dmnl

effect of energy efficiency on space heating=
1/effect of combined fabric insulation and energy efficiency standard on dwelling energy efficiency
Units: Dmnl

effect of energy efficiency standard improvement on dwelling energy efficiency
= WITH LOOKUP (
SAP rating/normal SAP value,
([(0,0)-(1,1)],(0.18,0),(0.23,0.048),(0.303,0.084),(0.365,0.154),(0.402,0.222
),(0.434,0.312),(0.455,0.407),(0.481,0.544),(0.507645,0.657895),(1,1)))
Units: Dmnl

effect of energy efficiency standard on cooking energy=
IF THEN ELSE(Time<2011, 1/(1+(effect of energy efficiency standard improvement on dwelling energy efficiency
)), 1/(1+(effect of energy efficiency standard improvement on dwelling energy efficiency
)+RAMP(0.01,2010,2050)))
Units: Dmnl

effect of energy efficiency standard on lighting and appliances energy=
IF THEN ELSE(Time<2011, 1/(1+(effect of energy efficiency standard improvement on dwelling energy efficiency
)), 1/(1+(
effect of energy efficiency standard improvement on dwelling energy efficiency
)+RAMP(0.01,2010,2050)))
Units: Dmnl

effect of fabric insulation on energy efficiency= WITH LOOKUP (
insulation factor/normal insulation,
([(0,0)-(1,1)],(0,0),(0.1,0.058),(0.2,0.162),(0.3,0.248),(0.4,0.305),(0.5
,0.364),(0.6,0.391),(0.7,0.491),(0.75,0.524)))
Units: Dmnl

effect of unfavourable climatic effects on international energy prices=
2*unfavourable climatic effects
Units: Dmnl

energy to carbon conversion for space heating=
Space Heating Energy Consumption*energy to carbon conversion factor
Units: Tonnes/Year

energy to carbon conversion appliances=
Appliances Energy Consumption*energy to carbon conversion factor
Units: Tonnes/Year

energy to carbon conversion ck=
Cooking Energy Consumption*energy to carbon conversion factor
Units: Tonnes/Year

energy to carbon conversion factor=
0.5246
Units: Dmnl [-1,1,0.01]

energy to carbon conversion hw=
Hot Water Energy Consumption*energy to carbon conversion factor
Units: Tonnes/Year

energy to carbon conversion l=

Lighting Energy Consumption*energy to carbon conversion factor
Units: Tonnes/Year

external air temp:=
GET XLS DATA ('Datasets.xlsx' , 'Datasets' , 'A' , 'B2')
Units: Deg Cent

frame factor=
0.7
Units: Dmnl [0.7,0.8,0.01]
frame factor for windows and doors (fraction of opening that is glazed) (0.7-0.8)

great discomfort=
LOOKUP EXTRAPOLATE(great discomfort lookup , dwelling int temp)
Units: Dmnl

great discomfort lookup(
[(30,33)-(40,43)],(31.1927,33.0877),(32.8746,33.7895),(34.4342,34.886),(35.9939
,36.2018),(37.5229,37.4298),(38.7156,38.5702),(39.9388,39.7544))
Units: Dmnl

growth in occupants activity level=
1.25
Units: Dmnl [0,3,0.1]

heat losses=
-40*average household size
Units: Watts

Hot Water Energy Consumption= INTEG (
(rate of hot water energy usage-energy to carbon conversion hw),
initial hot water usage)
Units: MWh

hot water energy consumption for UK housing stock=
Hot Water Energy Consumption*households/10^6
Units: TWh

hot water usage demand=
Occupants Comfort
Units: Dmnl

hot water usage energy:=
GET XLS DATA ('Datasets.xlsx' , 'Datasets' , 'A' , 'J2')
Units: MWh/Year

households=
-3.436e+008+182058*Time+0.067*Population
Units: Households

humidex value=
IF THEN ELSE(dwelling int temp<25
:OR:
relative humidity<50,
MAX(dwelling int temp*2.35,dwelling internal temp),
no discomfort from heat stress
)
Units: Dmnl

hw carbon depletion=
Carbon Emissions due to Hot Water Usage*carbon depletion factor
Units: Tonnes/Year

initial appliances usage=
1.07
Units: MWh

initial appliances usage carbon emissions=
0.375
Units: Tonnes

initial cooking carbon emissions=
0.55
Units: Tonnes

initial cooking energy=
1.36
Units: MWh

initial hot water usage=
6.64
Units: MWh

initial hot water usage carbon emissions=
2.7
Units: Tonnes

initial lighting usage=
0.55
Units: MWh

initial lighting usage carbon emissions=
0.21
Units: Tonnes

initial occupants comfort=
10
Units: com

initial occupants metabolic buildup=
0.9
Units: met

initial population=
5.5632e+007
Units: People

initial space heating carbon emissions=
5.85
Units: Tonnes

initial space heating energy=
13.18
Units: MWh

insulation factor=
0
Units: Watts/Year/Deg Cent [0,100,25]

l carbon depletion=
Carbon Emissions due to Lighting Energy*carbon depletion factor
Units: Tonnes/Year

lighting energy:=
GET XLS DATA ('Datasets.xlsx' , 'Datasets' , 'A' , 'L2')
Units: MWh/Year

Lighting Energy Consumption= INTEG (
(rate of lighting energy usage-energy to carbon conversion l),
initial lighting usage)
Units: MWh

lighting energy consumption for UK housing stock=
Lighting Energy Consumption*households/10^6
Units: TWh

lighting energy demand=
2
Units: Dmnl

mortality=
mortality lookup(average life expectancy/one year)
Units: 1/Year

mortality lookup(
[(20,0)-(80,0.006)],(20,0.00567),(30,0.00366),(40,0.00243),(50,0.00155),(60,0.00082),(70,0.00023),(80,0.0001))
Units: 1/Year

Natural heat transfer=
"discrepancy in int & ext temp"*(100-insulation factor)
Units: Watts/Year

no discomfort=
LOOKUP EXTRAPOLATE(no discomfort lookup, dwelling int temp)
Units: Dmnl

no discomfort from heat stress=
IF THEN ELSE(dwelling int temp<30
:AND:
relative humidity>25,
no discomfort,
some discomfort heat stress)
Units: Dmnl

no discomfort lookup(
[(20,26)-(25,30)],(0.202446,31.6009),(0.234251,29.2807),(0.292966,26.7588),(0.346789,25.2456),(0.449541,23.2281),(0.540061,22.1184),(0.657492,21.0088),(0.784709,20.5044),(0.899694,20.4035),(0.992661,20.3026),(20.0153,26.0526),(21.1009,26.3333),(22.0489,26.5439),(22.844,26.7368),(23.578,27.0351),(24.1896,27.3333),(25,27.8246))
Units: Dmnl

normal energy bills=
100
Units: Dmnl

normal insulation=
100
Units: Dmnl

normal SAP value=
100
Units: Dmnl

Occupants activity level=
(Occupants Metabolic Buildup*growth in occupants activity level)
Units: act

occupants behaviour=
2
Units: Dmnl [1,3,1]

Occupants Comfort= INTEG (
Perceived dwelling temp*"discrepancy in int & setpoint temp",
initial occupants comfort)
Units: Dmnl

Occupants Metabolic Buildup= INTEG (
((Perceived dwelling temp/30)/Occupants activity level)*0.05,
initial occupants metabolic buildup)
Units: Dmnl

one year=
1
Units: Year

Perceived dwelling temp=
IF THEN ELSE(humidex value<=(Occupants Comfort+Occupants Metabolic Buildup
)*probability of putting on clothing*probability of window opening
, (Occupants Comfort+Occupants Metabolic Buildup)*probability of putting on clothing
*probability of window opening , FORECAST
(humidex value,200,-100))
Units: Deg Cent

pi=
3.142
Units: Dmnl

```
Population= INTEG (
        births-deaths,
                initial population)
Units: People

population equilibrium time=
        2500
Units: Year

probability of putting on clothing=
        IF THEN ELSE(Occupants Metabolic Buildup>0
:AND:
some discomfort heat stress
>0,
                        LOOKUP EXTRAPOLATE (putting on clothing lookup,
Occupants Comfort/10), 0)
Units: Dmnl

probability of window opening=
        IF THEN ELSE(Occupants Metabolic Buildup>0
:AND:
some discomfort heat stress
>0,
                        LOOKUP EXTRAPOLATE (window opening lookup,
Occupants Comfort/100), 0)
Units: Dmnl

putting on clothing lookup(
        [(0,0.6)-(1,1)],(0,1),(0.058104,0.898246),(0.165138,0.814035),(0.284404,0.742105
),(0.425076,0.7),(0.593272,0.670175),(0.761468,0.650877),(0.990826,0.635088
))
Units: Dmnl

rate of appliances energy usage=
appliances energy*effect of energy efficiency standard on lighting and appliances energy
/effect of energy bills on energy consumption
        /0.88-0.25*
        FORECAST(appliances energy/0.88, 39, 507)
```

Units: MWh/Year

rate of hot water energy usage=
(hot water usage energy
*effect of energy efficiency on hot water energy/effect of energy bills on energy consumption
/0.88-0.25*FORECAST (hot water usage energy
/1.88, 39, 175))*(0.6*setpoint temp)/dwelling internal temp
)

Units: MWh

rate of lighting energy usage=
lighting energy*effect of energy efficiency standard on lighting and appliances energy
/effect of energy bills on energy consumption
/0.88-0.25*
FORECAST (lighting energy/0.88, 39, 600)) ,

Units: MWh/Year

rate of space heating=
(space heating energy
*effect of energy efficiency on space heating
/effect of energy bills on energy consumption
*1.14-0.15*FORECAST (space heating energy
0.53, 39, 450))(0.6*setpoint temp)/dwelling internal temp
) ,
Units: MWh/Year

relative humidity:=
GET XLS DATA('Datasets.xlsx' , 'Datasets' , 'A' , 'D2')
Units: per cent [0,100,1]

reproductive time=
90
Units: Year

SAP rating:=
GET XLS DATA ('Datasets.xlsx' , 'Datasets' , 'A' , 'N2')
Units: Dmnl [0,100,1]

setpoint temp=
20
Units: Deg Cent [10,30,0.5]

sh carbon depletion=
Space Heating Carbon Emissions*carbon depletion factor
Units: Tonnes/Year

solar flux=
150
Units: Dmnl [0,500,1]
solar flux on the applicable surface (solar irradiance)

solar transmittance factor for glazing=
0.76
Units: Dmnl [0.5,0.9,0.01]

some discomfort=
LOOKUP EXTRAPOLATE(some discomfort lookup, dwelling int temp)
Units: Dmnl

some discomfort heat stress=
IF THEN ELSE(dwelling int temp<36
:OR:
relative humidity>50,
some discomfort,
great discomfort)
Units: Dmnl

some discomfort lookup(
[(25,27)-(30,33)],(25.2905,27.0263),(26.422,27.4211),(27.4159,27.8684),(28.4557,28.4737),(29.2355,29.0789),(29.9847,29.8947))
Units: Dmnl

Space Heating Carbon Emissions= INTEG (
(energy to carbon conversion-sh carbon depletion)*0.5246,
initial space heating carbon emissions)
Units: Tonnes

space heating demand=
Occupants Comfort/ one com
Units: Dmnl

space heating energy:=
GET XLS DATA ('Datasets.xlsx' , 'Datasets' , 'A' , 'I2')
Units: **undefined**

Space Heating Energy Consumption= INTEG (
(rate of space heating-energy to carbon conversion),
initial space heating energy)
Units: MWh

space heating energy consumption for UK housing stock=
Space Heating Energy Consumption*households/10^6
Units: TWh

temperature conversion factor=
225
Units: Watts/Deg Cent [0,1000,1]

total annual household carbon emissions=
average annual carbon emissions per household*households/10^6
Units: Million tonnes

total annual household energy consumption=
average annual energy consumption per household*households/10^6
Units: TWh

"total dwelling heat gains (dhg)"=
(dhg due to appliances less cooking+dhg due to artificial lighting+dhg due to cooking
+dhg due to no of people+dhg due to water heating+average effect of solar gains
+heat losses)
Units: Watts

total floor area=
85
Units: **undefined**

unfavourable climatic effects=
SMOOTH ((1-(1/average annual carbon emissions per household)), 5)
Units: Dmnl

weighted average energy prices: =
GET XLS DATA ('Datasets.xlsx' , 'Datasets' , 'A' , 'Q2')
Units: **undefined**

window opening lookup (

[(0,0)-(1,1)],(0,0),(0.0795107,0.258772),(0.198777,0.495614),(0.40367,0.75
),(0.617737,0.890351),(0.801223,0.960526),(1,1))
Units: **undefined**

Appendix B
Expanded Simulation Results

See Tables B.1 and B.2.

Table B.1 Baseline household energy consumption (TWh)

Time (year)	Space heating	Hot water	Cooking	Lighting	Appliances
1970	247.54	124.71	25.54	10.33	20.10
1971	251.94	124.45	25.51	10.46	20.62
1972	244.24	124.89	25.37	10.66	21.61
1973	244.59	124.38	25.19	10.93	22.97
1974	252.12	123.99	25.06	11.22	24.67
1975	257.69	123.57	24.94	11.52	26.59
1976	254.39	122.74	24.79	11.83	28.40
1977	248.98	122.95	24.58	12.15	29.99
1978	253.92	122.98	24.36	12.46	31.37
1979	262.15	121.78	24.10	12.72	32.66
1980	284.08	120.88	23.84	13.02	33.84
1981	284.70	119.29	23.51	13.32	34.97
1982	283.88	118.27	23.15	13.58	36.13
1983	280.52	117.06	22.77	13.85	37.32
1984	277.80	115.53	22.31	14.08	38.58
1985	269.80	113.66	21.78	14.29	39.98
1986	290.58	112.90	21.31	14.54	41.72
1987	309.69	112.73	20.74	14.76	43.46
1988	318.24	110.91	20.13	14.96	45.20
1989	315.37	109.53	19.51	15.15	46.66
1990	300.92	108.20	18.88	15.29	47.93
1991	297.64	106.92	18.12	15.40	48.89
1992	321.25	105.65	17.51	15.55	49.70
1993	328.89	104.29	17.03	15.68	50.50

(continued)

M. G. Oladokun and C. O. Aigbavboa, *Simulation-Based Analysis of Energy and Carbon Emissions in the Housing Sector*, Green Energy and Technology,
https://doi.org/10.1007/978-3-319-75346-1

Table B.1 (continued)

Time (year)	Space heating	Hot water	Cooking	Lighting	Appliances
1994	341.94	103.20	16.67	15.85	51.28
1995	338.86	101.95	16.36	16.00	51.96
1996	330.04	100.77	16.10	16.13	52.46
1997	360.09	99.78	15.90	16.30	53.00
1998	353.88	98.85	15.75	16.50	53.56
1999	359.33	98.25	15.62	16.65	54.18
2000	362.70	97.02	15.52	16.84	54.77
2001	368.84	96.26	15.42	17.04	55.29
2002	380.81	94.99	15.34	17.25	55.84
2003	379.56	94.23	15.43	17.45	56.74
2004	382.72	93.86	15.48	17.44	57.87
2005	389.48	93.54	15.52	17.37	59.32
2006	383.08	92.46	15.48	17.17	60.82
2007	372.20	91.32	15.46	17.18	62.55
2008	356.90	89.82	15.38	17.11	63.65
2009	356.29	87.65	15.08	16.91	63.90
2010	339.89	84.85	14.02	16.10	64.57
2011	367.04	82.69	13.48	15.77	65.14
2012	320.58	81.25	11.95	15.67	65.65
2013	299.39	80.86	11.22	15.68	66.15
2014	290.28	80.97	10.86	14.81	67.28
2015	286.96	81.33	10.69	14.35	67.45
2016	285.51	81.40	10.61	14.09	67.17
2017	283.53	81.08	10.56	13.92	66.68
2018	281.35	80.58	10.54	13.80	66.10
2019	279.11	80.01	10.53	13.71	65.49
2020	276.88	79.41	10.52	13.62	64.87
2021	274.69	78.81	10.52	13.54	64.25
2022	272.54	78.21	10.51	13.46	63.64
2023	270.44	77.62	10.51	13.39	63.05
2024	268.39	77.04	10.50	13.31	62.46
2025	266.38	76.46	10.50	13.24	61.89
2026	264.41	75.90	10.50	13.16	61.33
2027	262.47	75.34	10.49	13.09	60.78
2028	260.57	74.78	10.49	13.01	60.24
2029	258.70	74.24	10.49	12.94	59.71
2030	256.86	73.70	10.48	12.86	59.19
2031	255.05	73.16	10.48	12.79	58.69
2032	253.26	72.63	10.47	12.72	58.19
2033	251.50	72.10	10.47	12.64	57.70

(continued)

Table B.1 (continued)

Time (year)	Space heating	Hot water	Cooking	Lighting	Appliances
2034	249.76	71.58	10.47	12.57	57.21
2035	248.04	71.06	10.46	12.50	56.74
2036	246.34	70.55	10.46	12.43	56.28
2037	244.66	70.04	10.46	12.35	55.82
2038	243.01	69.53	10.45	12.28	55.37
2039	241.37	69.03	10.45	12.21	54.92
2040	239.75	68.53	10.44	12.14	54.49
2041	238.15	68.03	10.44	12.07	54.05
2042	236.57	67.54	10.44	12.00	53.63
2043	235.01	67.05	10.43	11.93	53.21
2044	233.46	66.56	10.43	11.86	52.80
2045	231.93	66.08	10.42	11.79	52.39
2046	230.42	65.60	10.42	11.72	51.99
2047	228.93	65.13	10.42	11.65	51.59
2048	227.45	64.66	10.41	11.58	51.20
2049	225.99	64.19	10.41	11.51	50.81
2050	224.55	63.73	10.40	11.44	50.43

Table B.2 Baseline household carbon emissions (million Tonnes of CO_2)

Time (year)	Space heating	Hot water	Cooking	Lighting	Appliances
1970	109.87	3.94	10.33	3.94	7.04
1972	100.65	4.17	10.26	4.17	8.12
1973	96.85	4.25	10.21	4.25	8.51
1974	94.72	4.35	10.14	4.35	9.00
1975	95.09	4.46	10.09	4.46	9.63
1976	95.65	4.57	10.04	4.57	10.35
1977	93.79	4.70	9.98	4.70	11.08
1978	90.59	4.82	9.90	4.82	11.74
1979	89.81	4.95	9.81	4.95	12.33
1980	90.69	5.06	9.71	5.06	12.87
1981	95.85	5.17	9.61	5.17	13.36
1982	96.40	5.29	9.49	5.29	13.82
1983	95.10	5.40	9.35	5.40	14.29
1984	92.52	5.51	9.20	5.51	14.76
1985	89.69	5.61	9.02	5.61	15.25
1986	85.30	5.69	8.82	5.69	15.79
1987	88.41	5.79	8.63	5.79	16.44
1988	93.33	5.88	8.41	5.88	17.12
1989	95.86	5.96	8.17	5.96	17.81

(continued)

Table B.2 (continued)

Time (year)	Space heating	Hot water	Cooking	Lighting	Appliances
1990	94.47	6.04	7.93	6.04	18.43
1991	88.63	6.10	7.68	6.10	18.97
1992	84.51	6.15	7.39	6.15	19.40
1993	88.36	6.20	7.13	6.20	19.76
1994	90.21	6.26	6.92	6.26	20.09
1995	92.92	6.32	6.75	6.32	20.40
1996	91.40	6.38	6.61	6.38	20.69
1997	86.97	6.44	6.50	6.44	20.91
1998	92.41	6.50	6.41	6.50	21.13
1999	90.80	6.57	6.34	6.57	21.35
2000	90.24	6.64	6.28	6.64	21.59
2001	89.41	6.71	6.23	6.71	21.82
2002	89.24	6.79	6.19	6.79	22.04
2003	90.87	6.87	6.16	6.87	22.26
2004	89.36	6.95	6.17	6.95	22.57
2005	88.10	6.97	6.19	6.97	22.98
2006	87.88	6.96	6.20	6.96	23.51
2007	84.32	6.90	6.20	6.90	24.08
2008	78.48	6.88	6.19	6.88	24.74
2009	70.68	6.86	6.17	6.86	25.25
2010	66.34	6.79	6.07	6.79	25.47
2011	58.66	6.55	5.75	6.55	25.72
2012	78.69	6.38	5.50	6.38	25.95
2013	57.58	6.30	5.00	6.30	26.17
2014	59.64	6.28	4.65	6.28	26.37
2015	55.94	6.03	4.44	6.03	26.75
2016	56.10	5.83	4.33	5.83	26.91
2017	55.26	5.69	4.27	5.69	26.88
2018	54.66	5.61	4.24	5.61	26.73
2019	53.90	5.55	4.22	5.55	26.53
2020	53.19	5.50	4.21	5.50	26.29
2021	52.45	5.46	4.21	5.46	26.05
2022	51.73	5.43	4.21	5.43	25.80
2023	51.01	5.40	4.20	5.40	25.55
2024	50.30	5.37	4.20	5.37	25.31
2025	49.59	5.33	4.20	5.33	25.07
2026	48.89	5.30	4.20	5.30	24.84
2027	48.19	5.27	4.20	5.27	24.61
2028	47.49	5.24	4.19	5.24	24.39

(continued)

Table B.2 (continued)

Time (year)	Space heating	Hot water	Cooking	Lighting	Appliances
2029	46.80	5.21	4.19	5.21	24.17
2030	46.11	5.18	4.19	5.18	23.96
2031	45.42	5.15	4.19	5.15	23.75
2032	44.74	5.12	4.19	5.12	23.54
2033	44.05	5.09	4.19	5.09	23.34
2034	43.37	5.06	4.18	5.06	23.14
2035	42.68	5.04	4.18	5.04	22.94
2036	42.00	5.01	4.18	5.01	22.75
2037	41.32	4.98	4.18	4.98	22.56
2038	40.64	4.95	4.18	4.95	22.38
2039	39.96	4.92	4.18	4.92	22.20
2040	39.28	4.89	4.17	4.89	22.02
2041	38.59	4.86	4.17	4.86	21.84
2042	37.91	4.83	4.17	4.83	21.67
2043	37.23	4.80	4.17	4.80	21.49
2044	36.55	4.78	4.17	4.78	21.32
2045	35.87	4.75	4.17	4.75	21.16
2046	35.19	4.72	4.16	4.72	20.99
2047	34.51	4.69	4.16	4.69	20.83
2048	33.82	4.66	4.16	4.66	20.67
2049	33.14	4.63	4.16	4.63	20.51
2050	32.46	4.61	4.16	4.61	20.35

Appendix C
Expanded Policy Analysis Results

C.1 Household Energy Consumption by End-Use for the 'Baseline' and 'Efficiency' Scenarios

See Figs. C.1, C.2, C.3, C.4 and C.5.

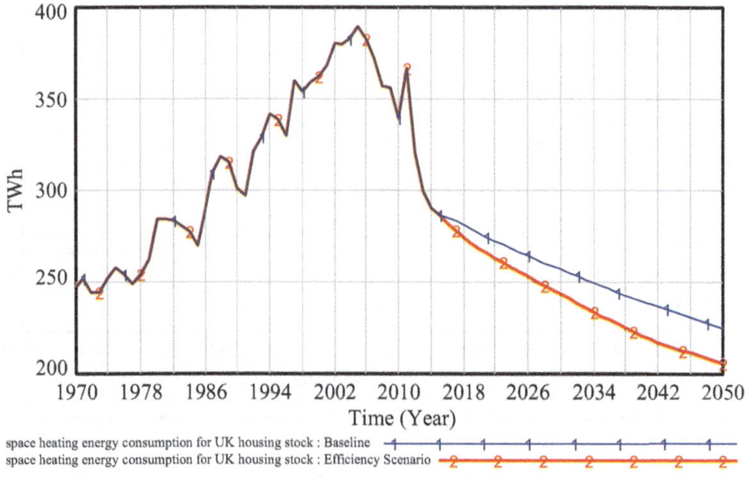

Fig. C.1 Space heating energy consumption for the 'baseline' and 'efficiency' scenarios

M. G. Oladokun and C. O. Aigbavboa, *Simulation-Based Analysis of Energy and Carbon Emissions in the Housing Sector*, Green Energy and Technology,
https://doi.org/10.1007/978-3-319-75346-1

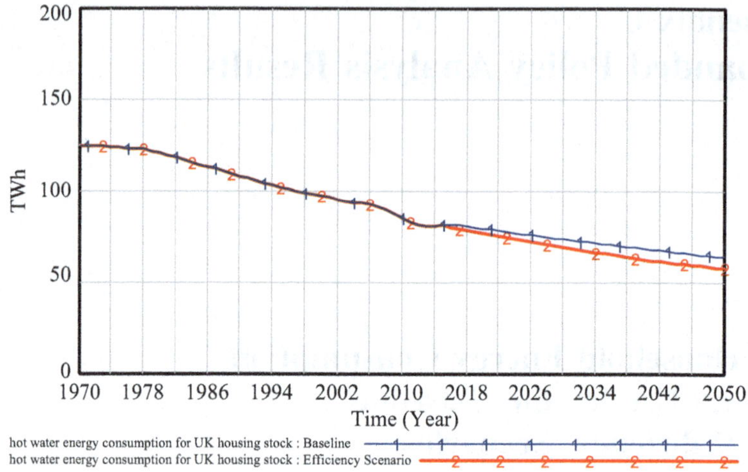

Fig. C.2 Hot water energy consumption for the 'baseline' and 'efficiency' scenarios

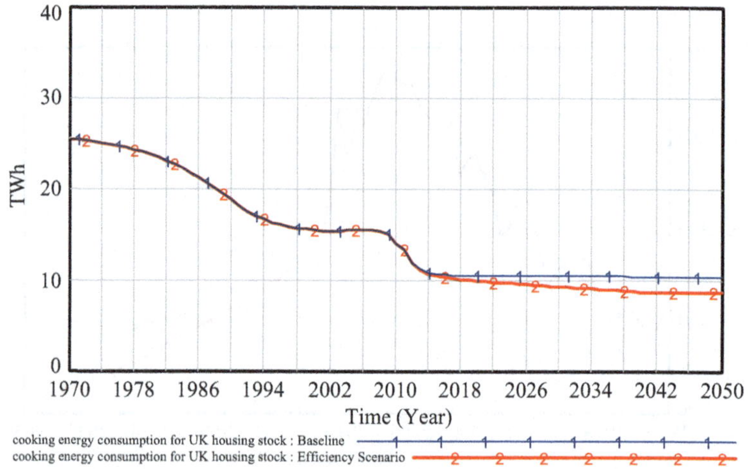

Fig. C.3 Cooking energy consumption for the 'baseline' and 'efficiency' scenarios

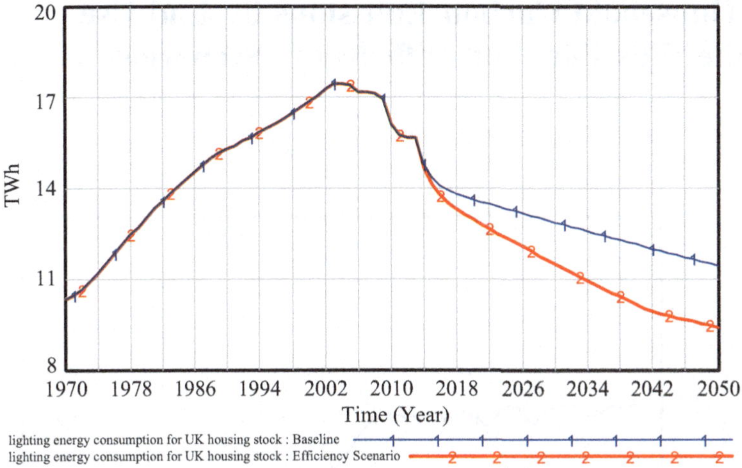

Fig. C.4 Lighting energy consumption for the 'baseline' and 'efficiency' scenarios

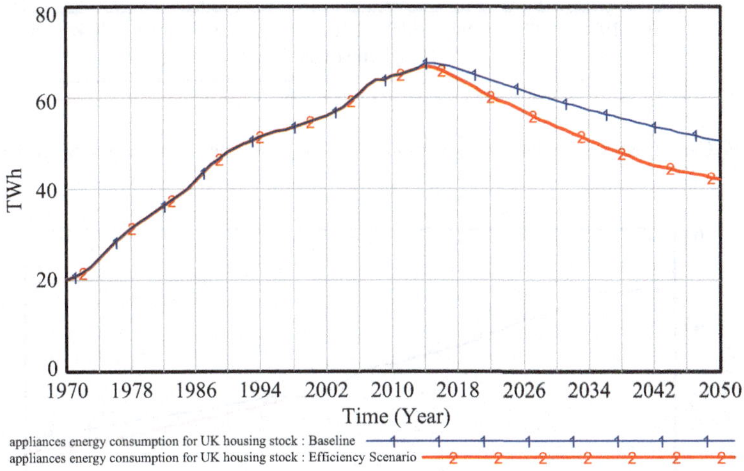

Fig. C.5 Appliance energy consumption for the 'baseline' and 'efficiency' scenarios

C.2 Household Carbon Emissions by End-Use for the 'Baseline' and 'Efficiency' Scenarios

See Figs. C.6, C.7, C.8, C.9 and C.10.

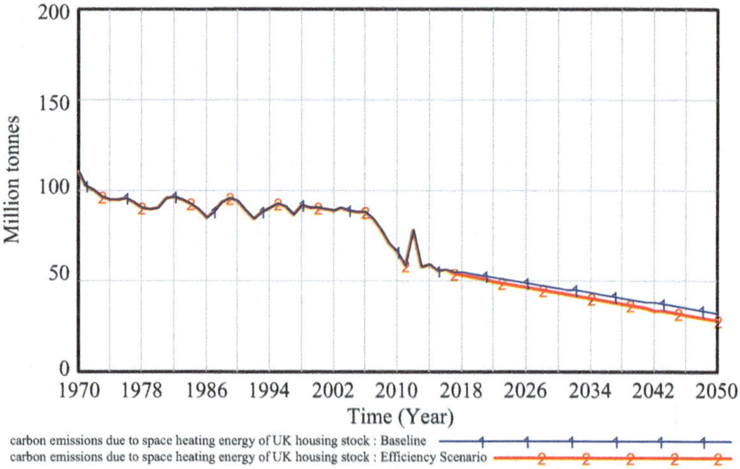

Fig. C.6 Carbon emissions due to space heating for the 'baseline' and 'efficiency' scenarios

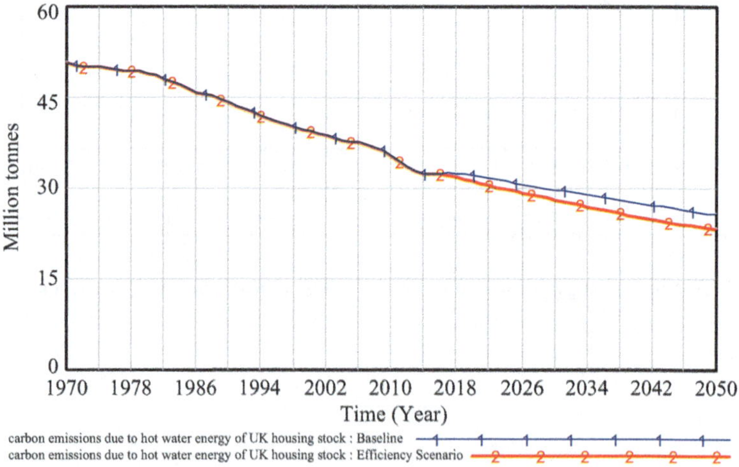

Fig. C.7 Carbon emissions due to hot water for the 'baseline' and 'efficiency' scenarios

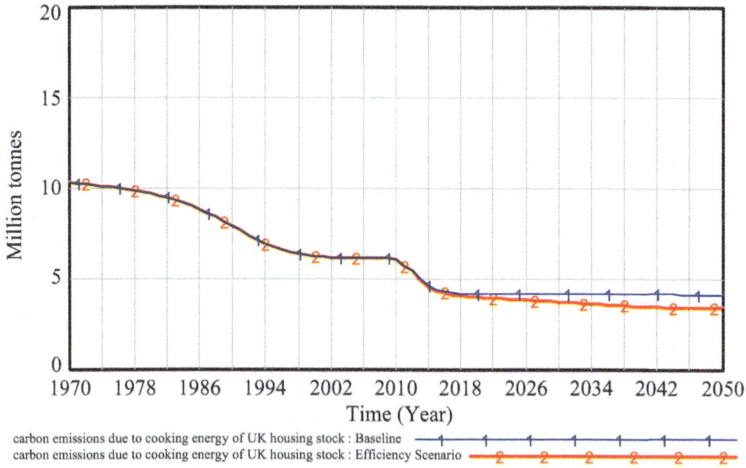

Fig. C.8 Carbon emissions due to cooking for the 'baseline' and 'efficiency' scenarios

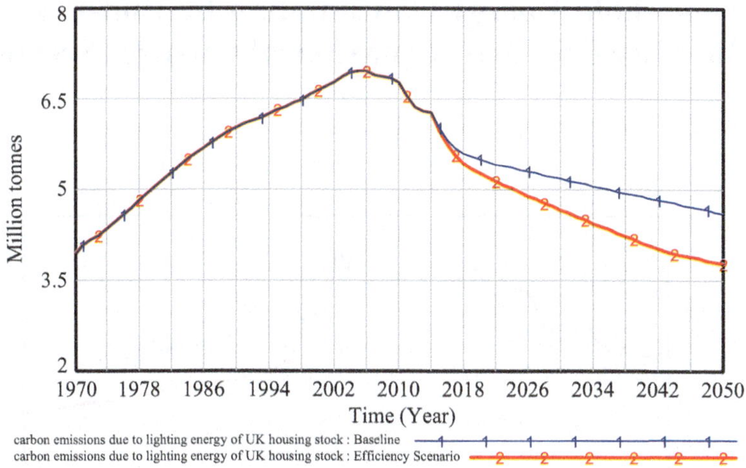

Fig. C.9 Carbon emissions due to lighting for the 'baseline' and 'efficiency' scenarios

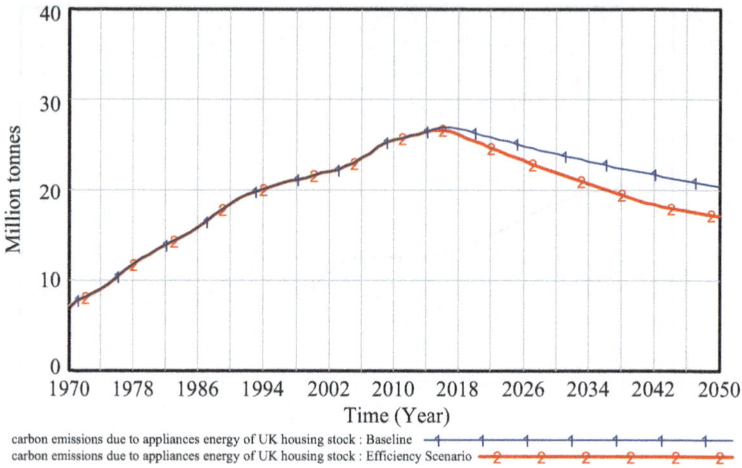

Fig. C.10 Carbon emissions due to appliances for the 'baseline' and 'efficiency' scenarios

C.3 Household Energy Consumption by End-Use for the 'Baseline' and 'Behavioural Change' Scenarios

See Figs. C.11, C.12, C.13, C.14 and C.15.

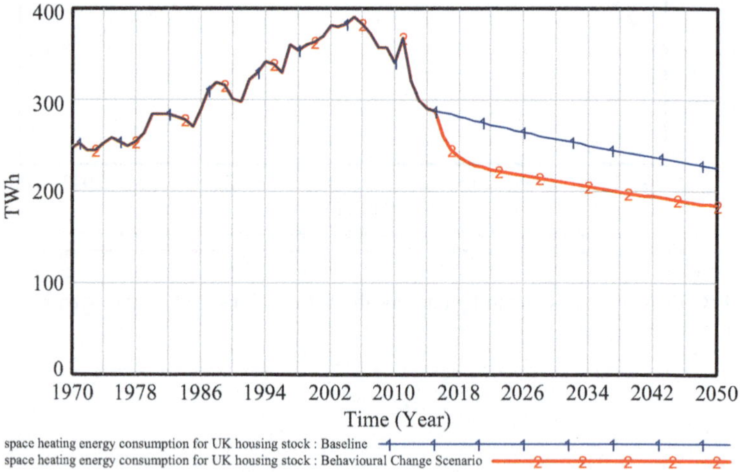

Fig. C.11 Space heating energy consumption for the 'baseline' and 'behavioural change' scenarios

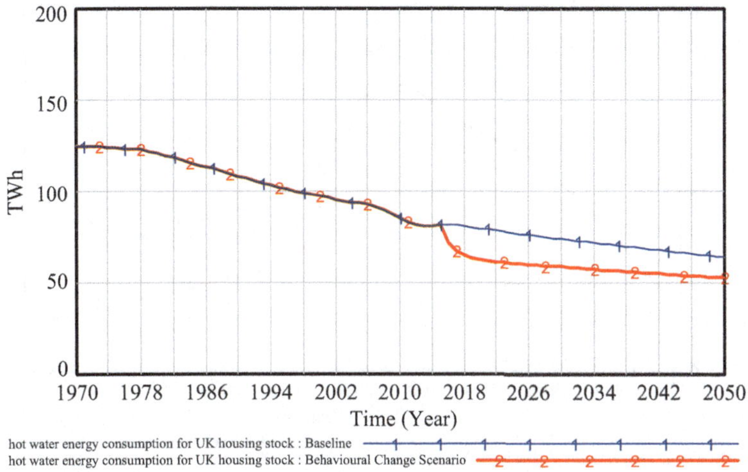

Fig. C.12 Hot water energy consumption for the 'baseline' and 'behavioural change' scenarios

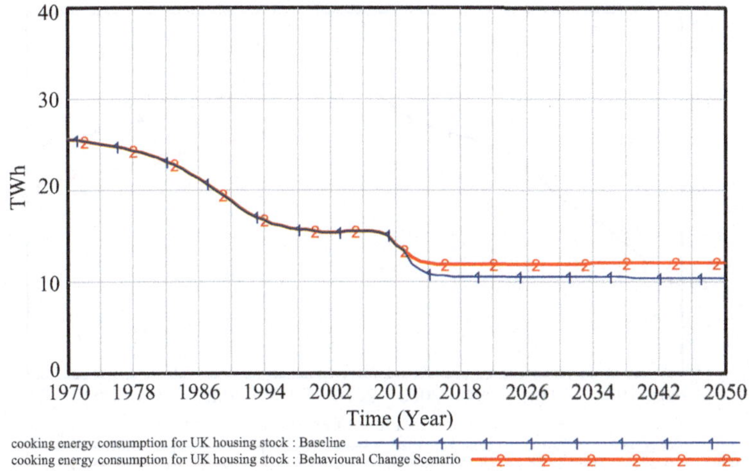

Fig. C.13 Cooking energy consumption for the 'baseline' and 'behavioural change' scenarios

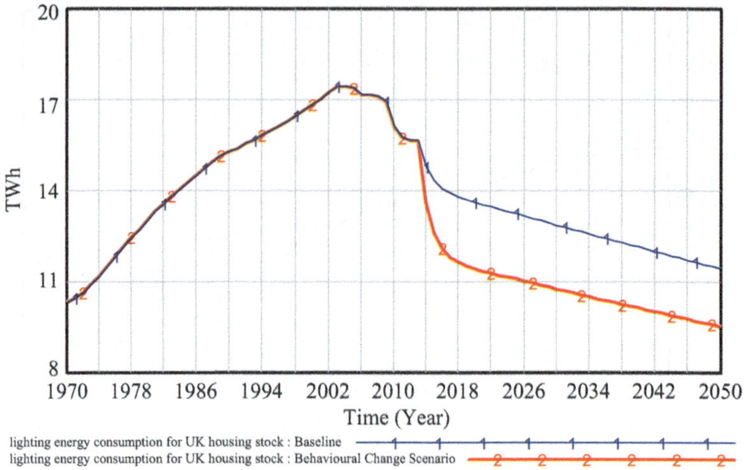

Fig. C.14 Lighting energy consumption for the 'baseline' and 'behavioural change' scenarios

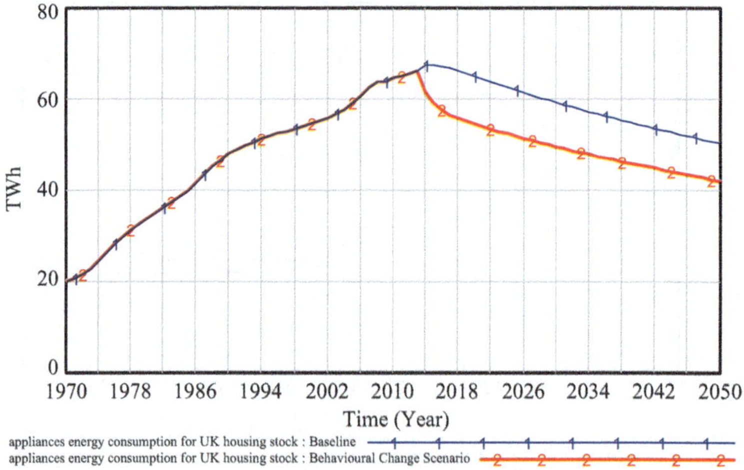

Fig. C.15 Appliance energy consumption for the 'baseline' and 'behavioural change' scenarios

C.4 Household Carbon Emissions by End-Use for the 'Baseline' and 'Behavioural Change' Scenarios

See Figs. C.16, C.17, C.18, C.19 and C.20.

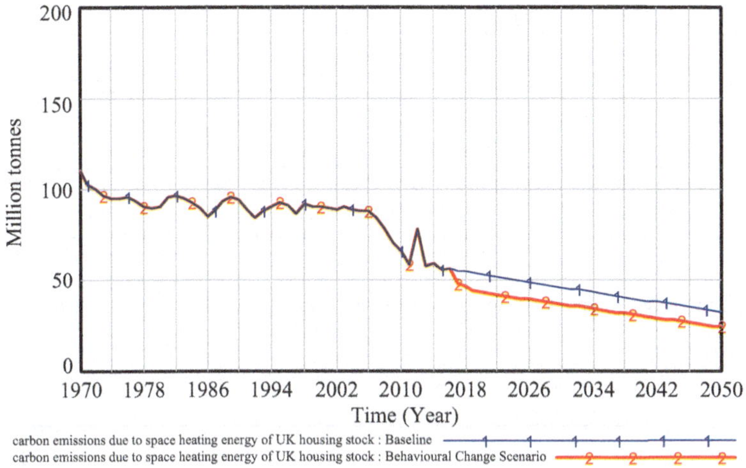

Fig. C.16 Carbon emissions due to space heating for the 'baseline' and 'behavioural change' scenarios

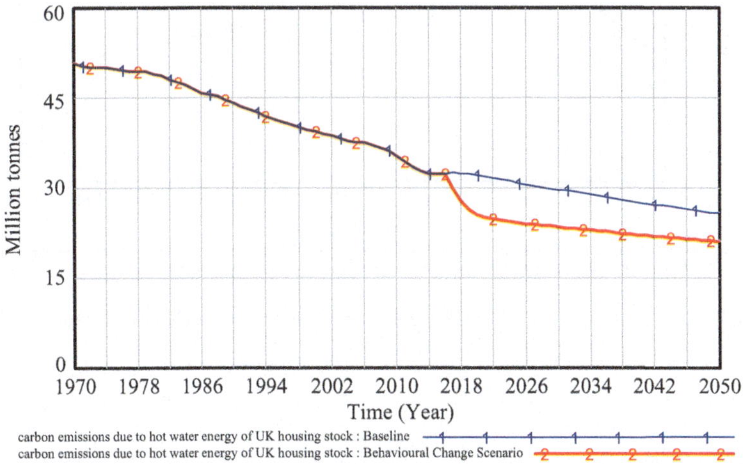

Fig. C.17 Carbon emissions due to hot water for the 'baseline' and 'behavioural change' scenarios

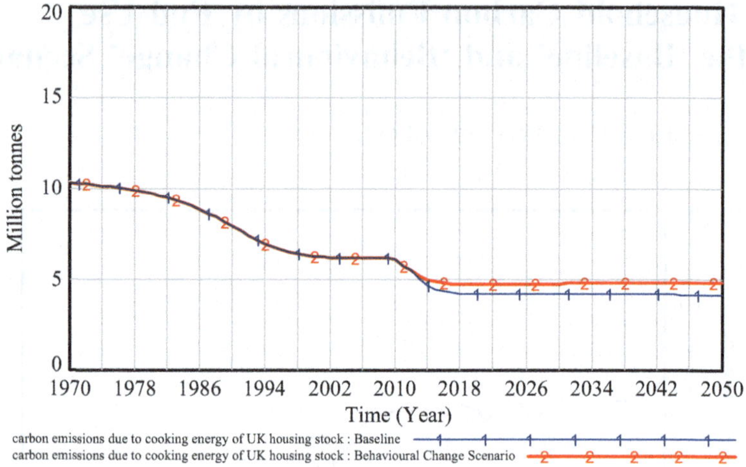

Fig. C.18 Carbon emissions due to cooking for the 'baseline' and 'behavioural change' scenarios

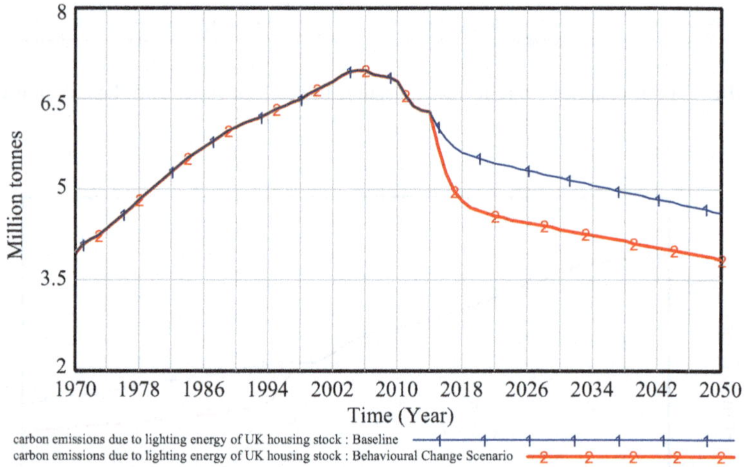

Fig. C.19 Carbon emissions due to lighting for the 'baseline' and 'behavioural change' scenarios

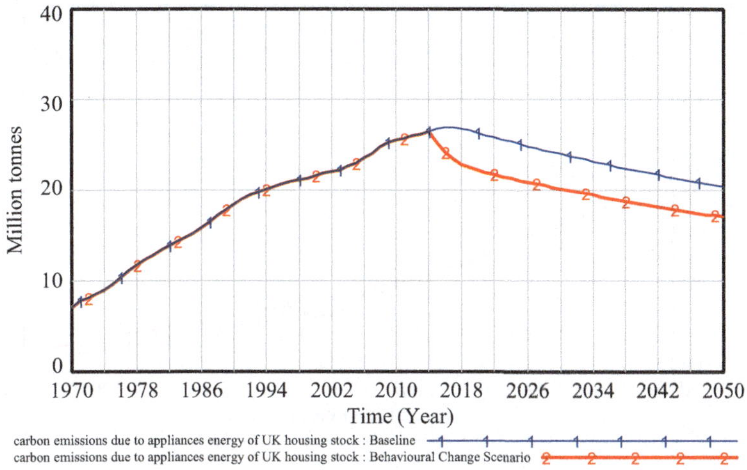

Fig. C.20 Carbon emissions due to appliances for the 'baseline' and 'behavioural change' scenarios

C.5 Household Energy Consumption by End-Use for the 'Baseline' and 'Economic' Scenarios

See Figs. C.21, C.22, C.23, C.24 and C.25.

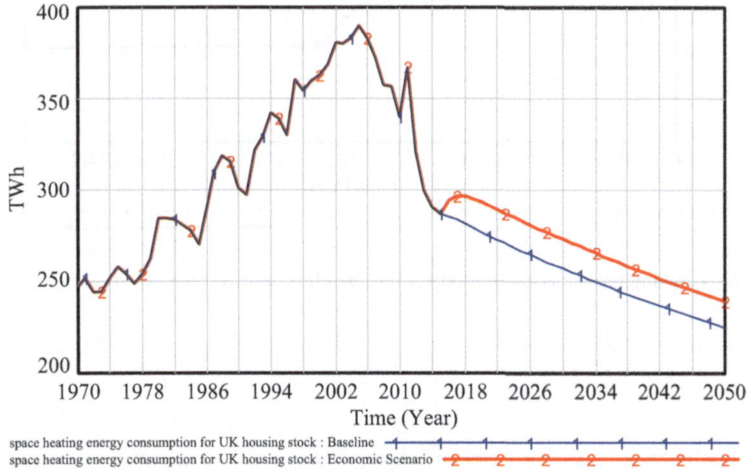

Fig. C.21 Space heating energy consumption for the 'baseline' and 'economic' scenarios

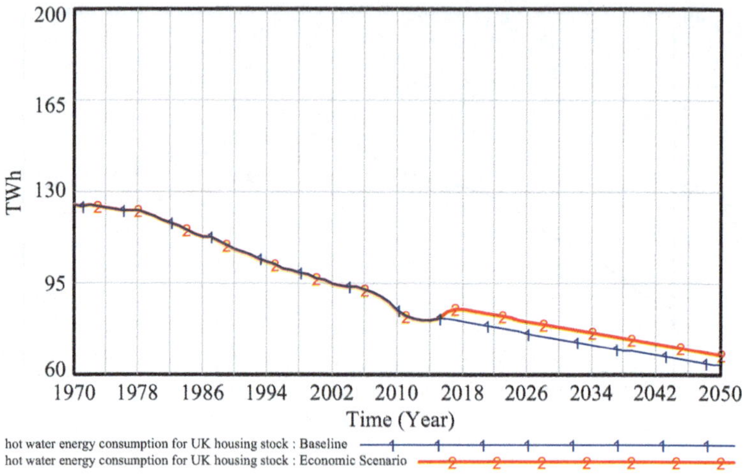

Fig. C.22 Hot water energy consumption for the 'baseline' and 'economic' scenarios

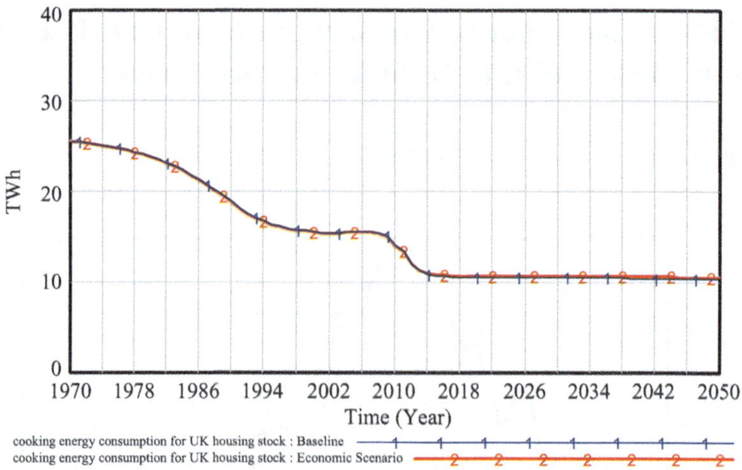

Fig. C.23 Cooking energy consumption for the 'baseline' and 'economic' scenarios

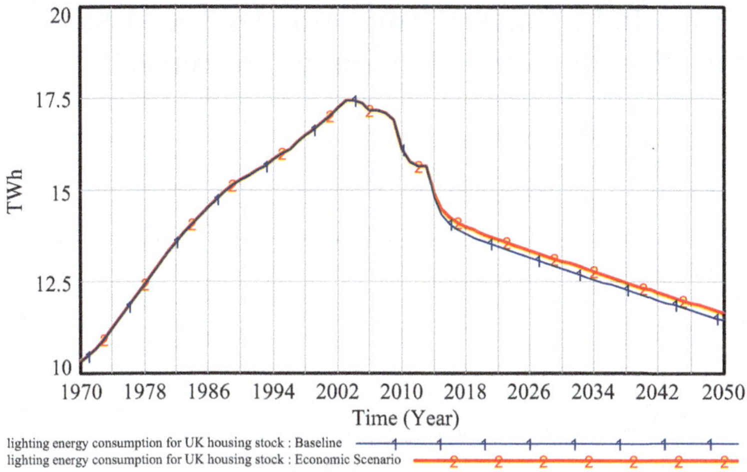

Fig. C.24 Lighting energy consumption for the 'baseline' and 'economic' scenarios

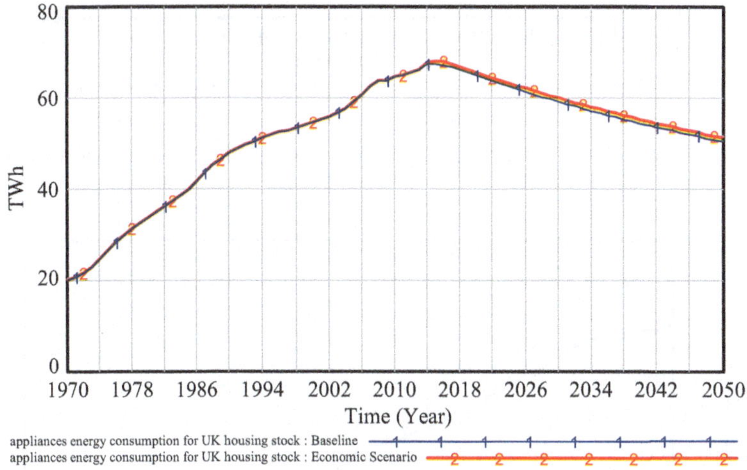

Fig. C.25 Appliance energy consumption for the 'baseline' and 'economic' scenarios

C.6 Household Carbon Emissions by End-Use for the 'Baseline' and 'Economic' Scenarios

See Figs. C.26, C.27, C.28, C.29 and C.30.

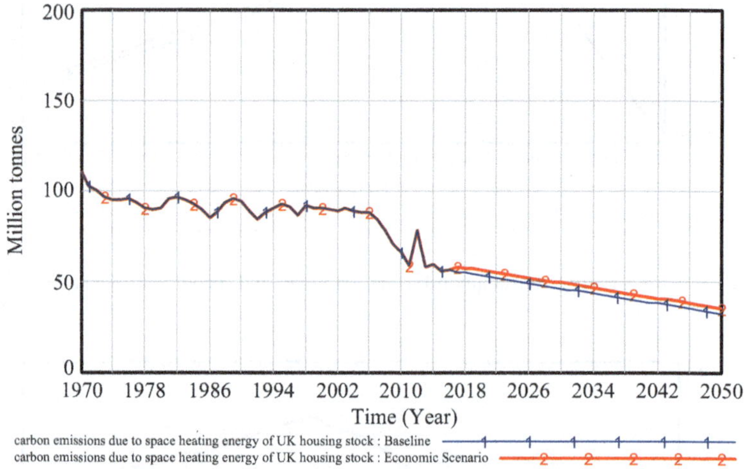

Fig. C.26 Carbon emissions due to space heating for the 'baseline' and 'economic' scenarios

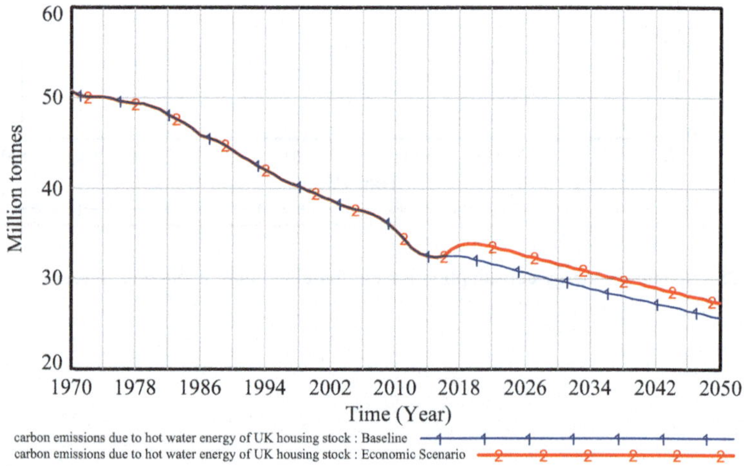

Fig. C.27 Carbon emissions due to hot water for the 'baseline' and 'economic' scenarios

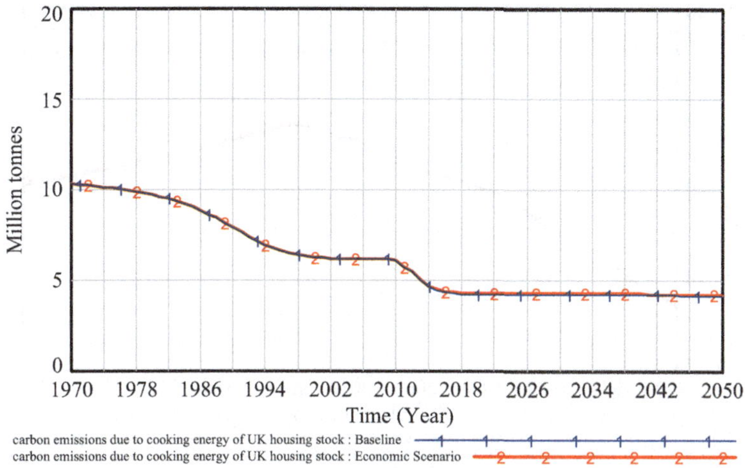

Fig. C.28 Carbon emissions due to cooking for the 'baseline' and 'economic' scenarios

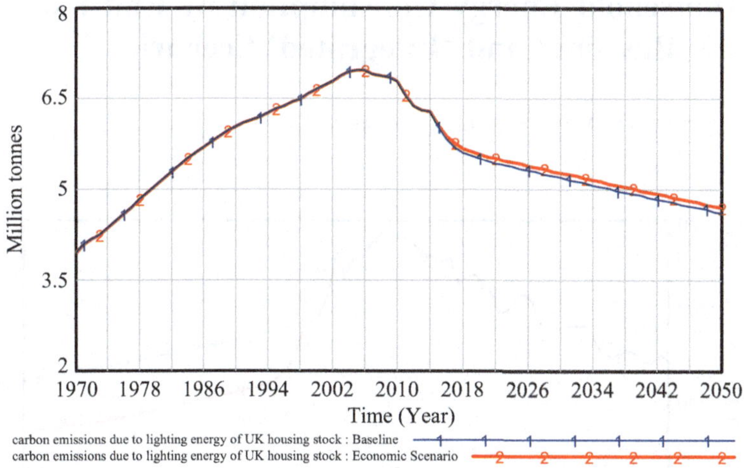

Fig. C.29 Carbon emissions due to lighting for the 'baseline' and 'economic' scenarios

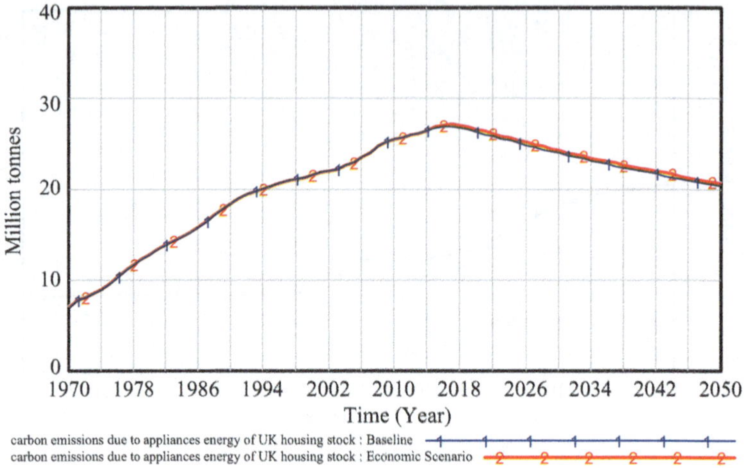

Fig. C.30 Carbon emissions due to appliances for the 'baseline' and 'economic' scenarios

C.7 Household Energy Consumption by End-Use for the 'Baseline' and 'Integrated' Scenarios

See Figs. C.31, C.32, C.33, C.34 and C.35.

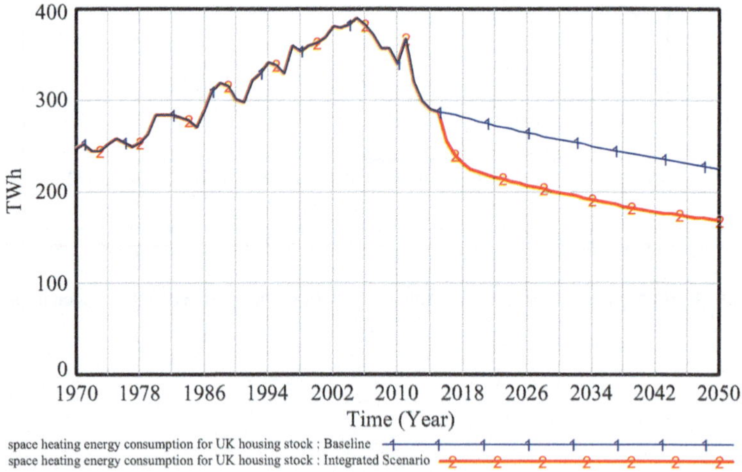

Fig. C.31 Space heating energy consumption for the 'baseline' and 'integrated' scenarios

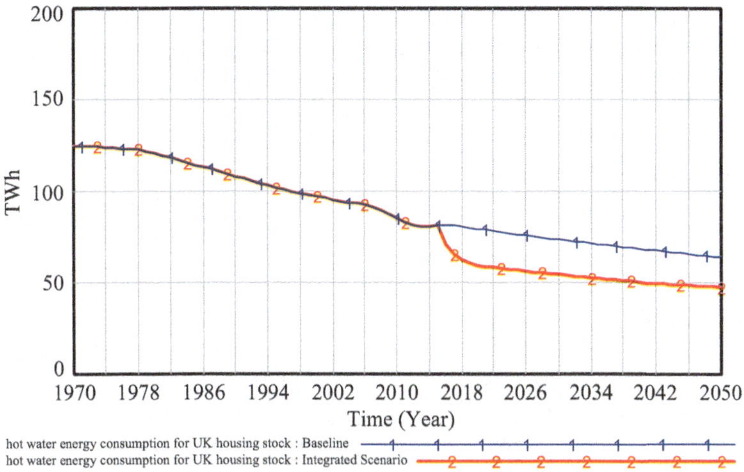

Fig. C.32 Hot water energy consumption for the 'baseline' and 'integrated' scenarios

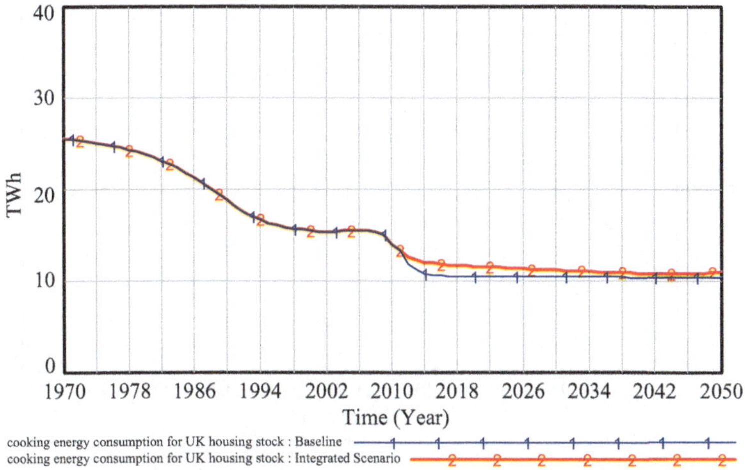

Fig. C.33 Cooking energy consumption for the 'baseline' and 'integrated' scenarios

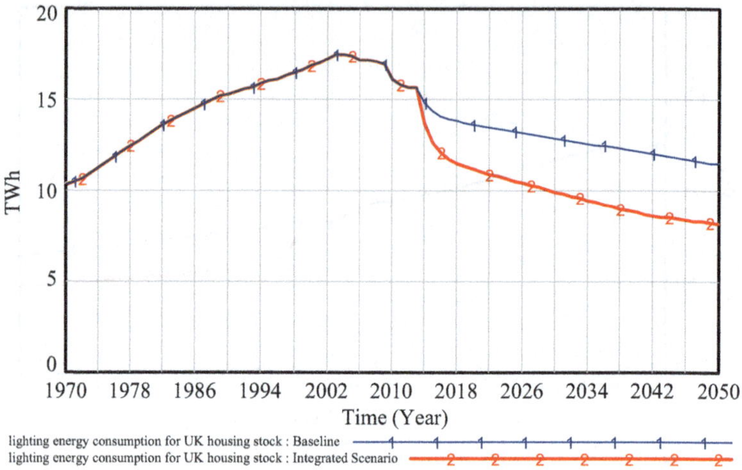

Fig. C.34 Lighting energy consumption for the 'baseline' and 'integrated' scenarios

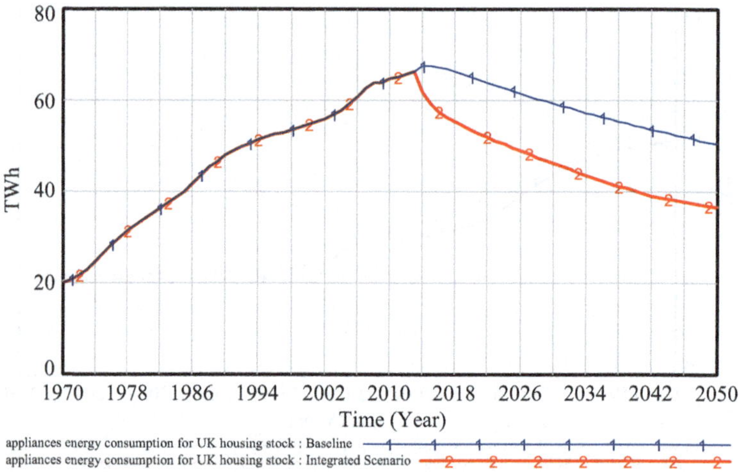

Fig. C.35 Appliance energy consumption for the 'baseline' and 'integrated' scenarios

C.8 Household Carbon Emissions by End-Use for the 'Baseline' and 'Integrated' Scenarios

See Figs. C.36, C.37, C.38, C.39 and C.40.

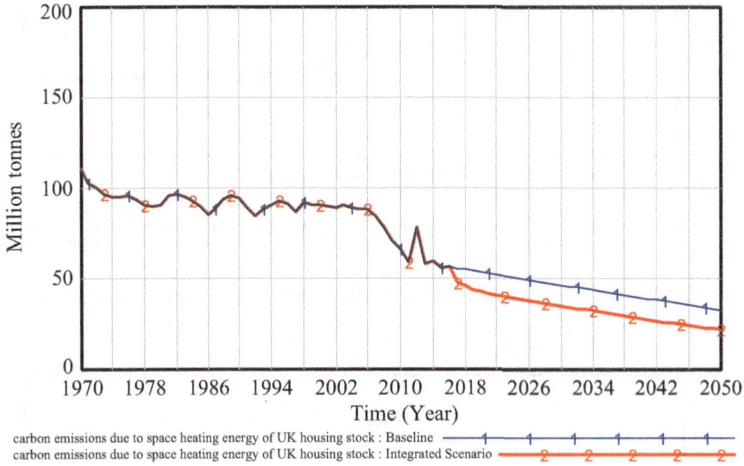

Fig. C.36 Carbon emissions due to space heating for the 'baseline' and 'integrated' scenarios

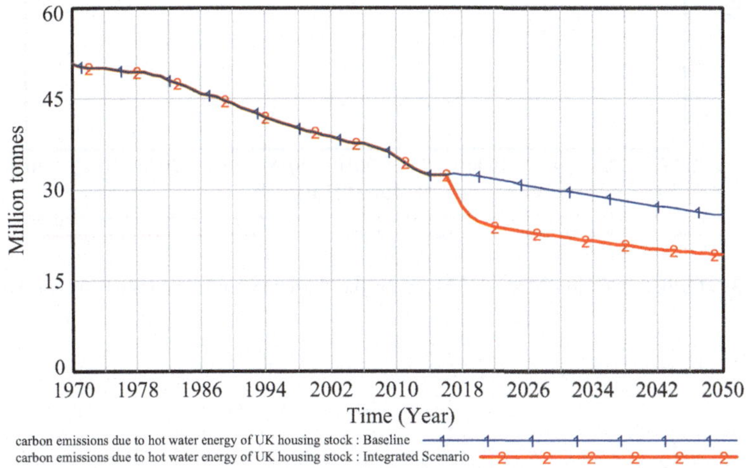

Fig. C.37 Carbon emissions due to hot water for the 'baseline' and 'integrated' scenarios

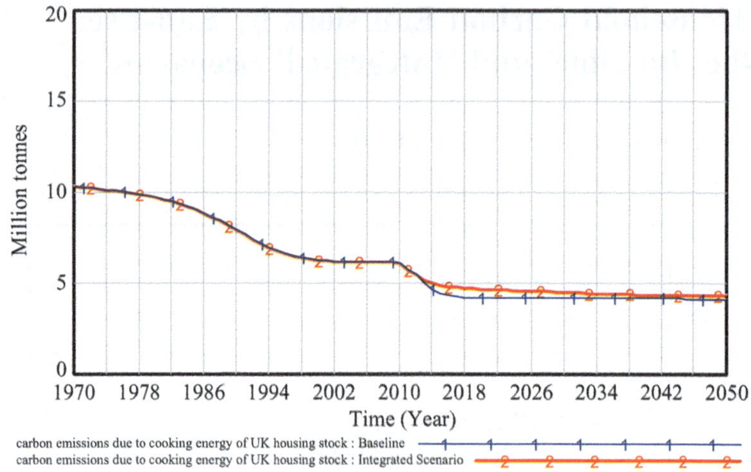

Fig. C.38 Carbon emissions due to cooking for the 'baseline' and 'integrated' scenarios

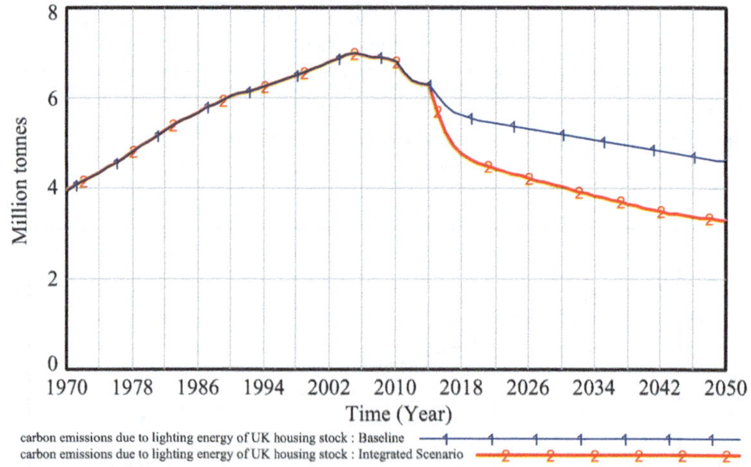

Fig. C.39 Carbon emissions due to lighting for the 'baseline' and 'integrated' scenarios

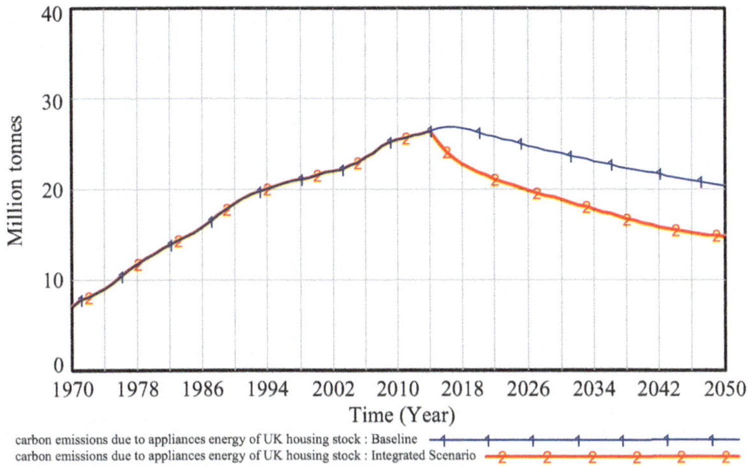

Fig. C.40 Carbon emissions due to appliances for the 'baseline' and 'integrated' scenarios

Appendix D
Expanded Validation Results

D.1 Time Step Changes for Euler Integration Method

See Tables D.1, D.2, D.3 and D.4.

Table D.1 Energy consumption per household—time step = 1

Time (year)	Space heating	Hot water	Cooking	Lighting	Appliances
1970	13.18	6.64	1.36	0.55	1.07
1971	13.28	6.56	1.34	0.55	1.09
1972	12.74	6.52	1.32	0.56	1.13
1973	12.64	6.43	1.30	0.56	1.19
1974	12.90	6.34	1.28	0.57	1.26
1975	13.05	6.26	1.26	0.58	1.35
1976	12.76	6.16	1.24	0.59	1.42
1977	12.37	6.11	1.22	0.60	1.49
1978	12.50	6.05	1.20	0.61	1.54
1979	12.78	5.94	1.18	0.62	1.59
1980	13.72	5.84	1.15	0.63	1.63
1981	13.63	5.71	1.13	0.64	1.67
1982	13.47	5.61	1.10	0.64	1.71
1983	13.19	5.50	1.07	0.65	1.75
1984	12.94	5.38	1.04	0.66	1.80
1985	12.46	5.25	1.01	0.66	1.85
1986	13.30	5.17	0.98	0.67	1.91
1987	14.05	5.11	0.94	0.67	1.97
1988	14.31	4.99	0.91	0.67	2.03
1989	14.06	4.88	0.87	0.68	2.08
1990	13.30	4.78	0.83	0.68	2.12
1991	13.05	4.69	0.79	0.67	2.14

(continued)

M. G. Oladokun and C. O. Aigbavboa, *Simulation-Based Analysis of Energy and Carbon Emissions in the Housing Sector*, Green Energy and Technology,
https://doi.org/10.1007/978-3-319-75346-1

Table D.1 (continued)

Time (year)	Space heating	Hot water	Cooking	Lighting	Appliances
1992	13.97	4.59	0.76	0.68	2.16
1993	14.18	4.50	0.73	0.68	2.18
1994	14.62	4.41	0.71	0.68	2.19
1995	14.37	4.32	0.69	0.68	2.20
1996	13.88	4.24	0.68	0.68	2.21
1997	15.03	4.16	0.66	0.68	2.21
1998	14.65	4.09	0.65	0.68	2.22
1999	14.76	4.04	0.64	0.68	2.23
2000	14.78	3.95	0.63	0.69	2.23
2001	14.91	3.89	0.62	0.69	2.24
2002	15.28	3.81	0.62	0.69	2.24
2003	15.11	3.75	0.61	0.69	2.26
2004	15.12	3.71	0.61	0.69	2.29
2005	15.27	3.67	0.61	0.68	2.33
2006	14.91	3.60	0.60	0.67	2.37
2007	14.38	3.53	0.60	0.66	2.42
2008	13.69	3.44	0.59	0.66	2.44
2009	13.56	3.34	0.57	0.64	2.43
2010	12.84	3.21	0.53	0.61	2.44
2011	13.77	3.10	0.51	0.59	2.44
2012	11.94	3.03	0.45	0.58	2.45
2013	11.07	2.99	0.41	0.58	2.45
2014	10.66	2.97	0.40	0.54	2.47
2015	10.46	2.96	0.39	0.52	2.46
2016	10.33	2.95	0.38	0.51	2.43
2017	10.19	2.91	0.38	0.50	2.40
2018	10.04	2.88	0.38	0.49	2.36
2019	9.89	2.84	0.37	0.49	2.32
2020	9.75	2.80	0.37	0.48	2.28
2021	9.60	2.76	0.37	0.47	2.25
2022	9.46	2.72	0.37	0.47	2.21
2023	9.33	2.68	0.36	0.46	2.17
2024	9.19	2.64	0.36	0.46	2.14
2025	9.07	2.60	0.36	0.45	2.11
2026	8.94	2.57	0.35	0.44	2.07
2027	8.82	2.53	0.35	0.44	2.04
2028	8.69	2.50	0.35	0.43	2.01
2029	8.58	2.46	0.35	0.43	1.98
2030	8.46	2.43	0.35	0.42	1.95
2031	8.35	2.39	0.34	0.42	1.92
2032	8.23	2.36	0.34	0.41	1.89

(continued)

Table D.1 (continued)

Time (year)	Space heating	Hot water	Cooking	Lighting	Appliances
2033	8.13	2.33	0.34	0.41	1.86
2034	8.02	2.30	0.34	0.40	1.84
2035	7.91	2.27	0.33	0.40	1.81
2036	7.81	2.24	0.33	0.39	1.78
2037	7.71	2.21	0.33	0.39	1.76
2038	7.61	2.18	0.33	0.38	1.73
2039	7.51	2.15	0.33	0.38	1.71
2040	7.42	2.12	0.32	0.38	1.69
2041	7.32	2.09	0.32	0.37	1.66
2042	7.23	2.06	0.32	0.37	1.64
2043	7.14	2.04	0.32	0.36	1.62
2044	7.05	2.01	0.31	0.36	1.59
2045	6.96	1.98	0.31	0.35	1.57
2046	6.88	1.96	0.31	0.35	1.55
2047	6.79	1.93	0.31	0.35	1.53
2048	6.71	1.91	0.31	0.34	1.51
2049	6.63	1.88	0.31	0.34	1.49
2050	6.55	1.86	0.30	0.33	1.47

Table D.2 Carbon emissions per household—time step = 1

Time (year)	Space heating	Hot water	Cooking energy	Lighting energy	Appliances
1970	5.85	2.70	0.55	0.21	0.38
1971	5.41	2.65	0.54	0.22	0.41
1972	5.25	2.61	0.54	0.22	0.42
1973	5.00	2.59	0.53	0.22	0.44
1974	4.85	2.56	0.52	0.22	0.46
1975	4.82	2.53	0.51	0.23	0.49
1976	4.80	2.49	0.50	0.23	0.52
1977	4.66	2.45	0.50	0.23	0.55
1978	4.46	2.43	0.49	0.24	0.58
1979	4.38	2.41	0.48	0.24	0.60
1980	4.38	2.37	0.47	0.24	0.62
1981	4.59	2.33	0.46	0.25	0.64
1982	4.57	2.28	0.45	0.25	0.66
1983	4.47	2.24	0.44	0.25	0.67
1984	4.31	2.20	0.43	0.26	0.69
1985	4.14	2.15	0.42	0.26	0.70

(continued)

Table D.2 (continued)

Time (year)	Space heating	Hot water	Cooking energy	Lighting energy	Appliances
1986	3.90	2.10	0.40	0.26	0.72
1987	4.01	2.06	0.39	0.26	0.75
1988	4.20	2.04	0.38	0.26	0.77
1989	4.27	1.99	0.36	0.27	0.79
1990	4.18	1.95	0.35	0.27	0.81
1991	3.89	1.91	0.34	0.27	0.83
1992	3.67	1.87	0.32	0.27	0.84
1993	3.81	1.84	0.31	0.27	0.85
1994	3.86	1.80	0.30	0.27	0.86
1995	3.94	1.76	0.29	0.27	0.87
1996	3.85	1.73	0.28	0.27	0.87
1997	3.63	1.69	0.27	0.27	0.87
1998	3.83	1.66	0.27	0.27	0.87
1999	3.73	1.63	0.26	0.27	0.88
2000	3.68	1.61	0.26	0.27	0.88
2001	3.62	1.58	0.25	0.27	0.88
2002	3.58	1.55	0.25	0.27	0.88
2003	3.62	1.52	0.25	0.27	0.89
2004	3.53	1.50	0.24	0.27	0.89
2005	3.45	1.48	0.24	0.27	0.90
2006	3.42	1.46	0.24	0.27	0.91
2007	3.26	1.44	0.24	0.27	0.93
2008	3.01	1.41	0.24	0.26	0.95
2009	2.69	1.38	0.23	0.26	0.96
2010	2.51	1.34	0.23	0.26	0.96
2011	2.20	1.29	0.22	0.25	0.96
2012	2.93	1.25	0.20	0.24	0.97
2013	2.13	1.21	0.18	0.23	0.97
2014	2.19	1.19	0.17	0.23	0.97
2015	2.04	1.18	0.16	0.22	0.98
2016	2.03	1.18	0.16	0.21	0.97
2017	1.99	1.17	0.15	0.20	0.97
2018	1.95	1.16	0.15	0.20	0.95
2019	1.91	1.14	0.15	0.20	0.94
2020	1.87	1.13	0.15	0.19	0.93
2021	1.83	1.11	0.15	0.19	0.91
2022	1.80	1.10	0.15	0.19	0.90
2023	1.76	1.08	0.14	0.19	0.88
2024	1.72	1.07	0.14	0.18	0.87

(continued)

Table D.2 (continued)

Time (year)	Space heating	Hot water	Cooking energy	Lighting energy	Appliances
2025	1.69	1.05	0.14	0.18	0.85
2026	1.65	1.04	0.14	0.18	0.84
2027	1.62	1.02	0.14	0.18	0.83
2028	1.58	1.01	0.14	0.17	0.81
2029	1.55	0.99	0.14	0.17	0.80
2030	1.52	0.98	0.14	0.17	0.79
2031	1.49	0.97	0.14	0.17	0.78
2032	1.45	0.95	0.14	0.17	0.77
2033	1.42	0.94	0.14	0.16	0.75
2034	1.39	0.93	0.13	0.16	0.74
2035	1.36	0.92	0.13	0.16	0.73
2036	1.33	0.90	0.13	0.16	0.72
2037	1.30	0.89	0.13	0.16	0.71
2038	1.27	0.88	0.13	0.15	0.70
2039	1.24	0.87	0.13	0.15	0.69
2040	1.22	0.86	0.13	0.15	0.68
2041	1.19	0.84	0.13	0.15	0.67
2042	1.16	0.83	0.13	0.15	0.66
2043	1.13	0.82	0.13	0.15	0.65
2044	1.10	0.81	0.13	0.14	0.64
2045	1.08	0.80	0.13	0.14	0.64
2046	1.05	0.79	0.12	0.14	0.63
2047	1.02	0.78	0.12	0.14	0.62
2048	1.00	0.77	0.12	0.14	0.61
2049	0.97	0.76	0.12	0.14	0.60
2050	0.95	0.75	0.12	0.13	0.59

Table D.3 Energy consumption per household—time step = 0.5

Time (year)	Space heating	Hot Water	Cooking	Lighting	Appliances
1970	13.18	6.64	1.36	0.55	1.07
1971	13.12	6.57	1.34	0.55	1.09
1972	12.76	6.51	1.32	0.56	1.14
1973	12.74	6.42	1.30	0.57	1.19
1974	12.93	6.34	1.28	0.57	1.27
1975	12.96	6.25	1.26	0.58	1.35
1976	12.69	6.17	1.24	0.59	1.42
1977	12.46	6.11	1.22	0.60	1.49
1978	12.59	6.04	1.20	0.61	1.54

(continued)

Table D.3 (continued)

Time (year)	Space heating	Hot Water	Cooking	Lighting	Appliances
1979	13.00	5.94	1.17	0.62	1.59
1980	13.58	5.83	1.15	0.63	1.63
1981	13.53	5.71	1.12	0.64	1.67
1982	13.38	5.61	1.10	0.64	1.71
1983	13.15	5.50	1.07	0.65	1.75
1984	12.86	5.38	1.04	0.66	1.80
1985	12.75	5.26	1.01	0.66	1.85
1986	13.43	5.18	0.97	0.67	1.91
1987	14.00	5.10	0.94	0.67	1.97
1988	14.16	4.99	0.90	0.67	2.03
1989	13.85	4.88	0.87	0.67	2.08
1990	13.32	4.79	0.83	0.68	2.11
1991	13.35	4.69	0.80	0.68	2.14
1992	13.95	4.59	0.76	0.68	2.16
1993	14.23	4.50	0.74	0.68	2.18
1994	14.48	4.41	0.71	0.68	2.19
1995	14.23	4.33	0.70	0.68	2.20
1996	14.22	4.24	0.68	0.68	2.21
1997	14.83	4.17	0.67	0.68	2.21
1998	14.66	4.10	0.65	0.68	2.22
1999	14.74	4.03	0.64	0.68	2.23
2000	14.80	3.96	0.63	0.69	2.23
2001	14.98	3.89	0.62	0.69	2.24
2002	15.18	3.81	0.62	0.69	2.24
2003	15.10	3.76	0.62	0.69	2.26
2004	15.15	3.71	0.61	0.69	2.29
2005	15.16	3.66	0.61	0.68	2.33
2006	14.81	3.60	0.60	0.67	2.37
2007	14.28	3.52	0.60	0.66	2.41
2008	13.78	3.44	0.59	0.65	2.43
2009	13.46	3.33	0.57	0.64	2.43
2010	13.20	3.21	0.53	0.61	2.44
2011	13.27	3.11	0.51	0.59	2.44
2012	11.91	3.04	0.45	0.59	2.44
2013	11.17	3.00	0.42	0.58	2.45
2014	10.76	2.98	0.41	0.55	2.46
2015	10.54	2.97	0.39	0.53	2.45
2016	10.34	2.94	0.39	0.51	2.42
2017	10.16	2.90	0.38	0.50	2.39
2018	10.00	2.86	0.38	0.49	2.36

(continued)

Table D.3 (continued)

Time (year)	Space heating	Hot Water	Cooking	Lighting	Appliances
2019	9.85	2.82	0.37	0.49	2.32
2020	9.70	2.78	0.37	0.48	2.28
2021	9.56	2.75	0.37	0.47	2.25
2022	9.43	2.71	0.37	0.47	2.21
2023	9.30	2.67	0.36	0.46	2.17
2024	9.17	2.63	0.36	0.46	2.14
2025	9.04	2.60	0.36	0.45	2.11
2026	8.92	2.56	0.35	0.44	2.07
2027	8.80	2.53	0.35	0.44	2.04
2028	8.68	2.49	0.35	0.43	2.01
2029	8.56	2.46	0.35	0.43	1.98
2030	8.45	2.43	0.35	0.42	1.95
2031	8.33	2.39	0.34	0.42	1.92
2032	8.22	2.36	0.34	0.41	1.89
2033	8.12	2.33	0.34	0.41	1.86
2034	8.01	2.30	0.34	0.40	1.84
2035	7.91	2.27	0.33	0.40	1.81
2036	7.80	2.24	0.33	0.39	1.79
2037	7.70	2.21	0.33	0.39	1.76
2038	7.61	2.18	0.33	0.38	1.73
2039	7.51	2.15	0.33	0.38	1.71
2040	7.41	2.12	0.32	0.38	1.69
2041	7.32	2.09	0.32	0.37	1.66
2042	7.23	2.06	0.32	0.37	1.64
2043	7.14	2.04	0.32	0.36	1.62
2044	7.05	2.01	0.31	0.36	1.60
2045	6.96	1.98	0.31	0.35	1.57
2046	6.88	1.96	0.31	0.35	1.55
2047	6.79	1.93	0.31	0.35	1.53
2048	6.71	1.91	0.31	0.34	1.51
2049	6.63	1.88	0.31	0.34	1.49
2050	6.55	1.86	0.30	0.33	1.47

Table D.4 Carbon emissions per household—time step = 0.5

Time (year)	Space heating	Hot water	Cooking	Lighting	Appliances
1970	5.85	2.70	0.55	0.21	0.38
1971	5.48	2.65	0.54	0.21	0.40
1972	5.23	2.62	0.53	0.22	0.42
1973	5.02	2.59	0.53	0.22	0.44
1974	4.90	2.56	0.52	0.22	0.46
1975	4.85	2.53	0.51	0.23	0.49
1976	4.77	2.49	0.50	0.23	0.52
1977	4.62	2.46	0.50	0.23	0.55
1978	4.49	2.43	0.49	0.24	0.58
1979	4.43	2.40	0.48	0.24	0.60
1980	4.49	2.37	0.47	0.24	0.62
1981	4.55	2.32	0.46	0.25	0.64
1982	4.52	2.28	0.45	0.25	0.65
1983	4.42	2.24	0.44	0.25	0.67
1984	4.28	2.20	0.43	0.26	0.69
1985	4.10	2.15	0.42	0.26	0.71
1986	4.03	2.10	0.40	0.26	0.73
1987	4.11	2.07	0.39	0.26	0.75
1988	4.20	2.03	0.38	0.26	0.77
1989	4.20	1.99	0.36	0.27	0.79
1990	4.06	1.95	0.35	0.27	0.81
1991	3.86	1.91	0.34	0.27	0.83
1992	3.81	1.87	0.32	0.27	0.84
1993	3.84	1.84	0.31	0.27	0.85
1994	3.89	1.80	0.30	0.27	0.86
1995	3.88	1.76	0.29	0.27	0.86
1996	3.76	1.73	0.28	0.27	0.87
1997	3.76	1.70	0.27	0.27	0.87
1998	3.77	1.66	0.27	0.27	0.87
1999	3.72	1.64	0.26	0.27	0.88
2000	3.67	1.61	0.26	0.27	0.88
2001	3.62	1.58	0.25	0.27	0.88
2002	3.61	1.55	0.25	0.27	0.88
2003	3.58	1.52	0.25	0.27	0.89
2004	3.52	1.50	0.24	0.27	0.89
2005	3.47	1.48	0.24	0.27	0.90
2006	3.37	1.46	0.24	0.27	0.92
2007	3.20	1.43	0.24	0.27	0.93
2008	2.95	1.41	0.24	0.26	0.95
2009	2.71	1.37	0.23	0.26	0.96

(continued)

Table D.4 (continued)

Time (year)	Space heating	Hot water	Cooking	Lighting	Appliances
2010	2.47	1.33	0.23	0.25	0.96
2011	2.36	1.29	0.21	0.25	0.96
2012	2.49	1.25	0.20	0.24	0.97
2013	2.30	1.22	0.19	0.23	0.97
2014	2.17	1.20	0.17	0.23	0.97
2015	2.09	1.19	0.17	0.22	0.97
2016	2.04	1.18	0.16	0.21	0.97
2017	1.99	1.17	0.16	0.21	0.96
2018	1.94	1.15	0.15	0.20	0.95
2019	1.90	1.14	0.15	0.20	0.94
2020	1.86	1.12	0.15	0.19	0.92
2021	1.83	1.11	0.15	0.19	0.91
2022	1.79	1.09	0.15	0.19	0.90
2023	1.75	1.08	0.15	0.19	0.88
2024	1.72	1.06	0.14	0.18	0.87
2025	1.68	1.05	0.14	0.18	0.85
2026	1.65	1.03	0.14	0.18	0.84
2027	1.61	1.02	0.14	0.18	0.83
2028	1.58	1.01	0.14	0.17	0.81
2029	1.55	0.99	0.14	0.17	0.80
2030	1.52	0.98	0.14	0.17	0.79
2031	1.48	0.97	0.14	0.17	0.78
2032	1.45	0.95	0.14	0.17	0.77
2033	1.42	0.94	0.14	0.16	0.75
2034	1.39	0.93	0.13	0.16	0.74
2035	1.36	0.92	0.13	0.16	0.73
2036	1.33	0.90	0.13	0.16	0.72
2037	1.30	0.89	0.13	0.16	0.71
2038	1.27	0.88	0.13	0.15	0.70
2039	1.24	0.87	0.13	0.15	0.69
2040	1.21	0.86	0.13	0.15	0.68
2041	1.19	0.84	0.13	0.15	0.67
2042	1.16	0.83	0.13	0.15	0.66
2043	1.13	0.82	0.13	0.15	0.65
2044	1.10	0.81	0.13	0.14	0.64
2045	1.08	0.80	0.13	0.14	0.64
2046	1.05	0.79	0.12	0.14	0.63
2047	1.02	0.78	0.12	0.14	0.62
2048	1.00	0.77	0.12	0.14	0.61
2049	0.97	0.76	0.12	0.14	0.60
2050	0.95	0.75	0.12	0.13	0.59

Index

M. G. Oladokun and C. O. Aigbavboa, *Simulation-Based Analysis of Energy and Carbon Emissions in the Housing Sector*, Green Energy and Technology,
https://doi.org/10.1007/978-3-319-75346-1

The manufacturer's authorised representative in the EU is Springer
Nature Customer Service Centre GmbH, Europaplatz 3, 69115 Heidelberg,
Germany. If you have any concerns regarding our products, please
contact ProductSafety@springernature.com

Printed and bound by CPI Group (UK) Ltd, Croydon, CR0 4YY
30/09/2026
02225662-0007